BEYOND EARTH

BEYOND EARTH

OUR PATH TO A NEW HOME IN THE PLANETS

CHARLES WOHLFORTH AND
AMANDA R. HENDRIX, PHD

PANTHEON BOOKS, NEW YORK

Pantheon Books and colophon are registered trademarks of Penguin
Random House LLC.

Library of Congress Cataloging-in-Publication Data
Names: Wohlforth, Charles P., Hendrix, Amanda R.
Title: Beyond Earth : our path to a new home in the planets /
Charles Wohlforth and Amanda R. Hendrix, Ph.D.
Description: New York : Pantheon Books, 2016. Includes index.
Identifiers: LCCN 2016018868 (print). LCCN 2016009498 (ebook).
ISBN 9780804197977 (hardcover) ISBN 9780804197984 (ebook).
Subjects: LCSH: Manned space flight. Astronautics. Space flight—
Physiological effect. Space flight—Psychological aspects.
Classification: LCC TL790 (print). LCC TL790 .W63 2016 (ebook).
DDC 629.45/5—dc23. LC record available at:
lccn.loc.gov/2016018868

www.pantheonbooks.com

Jacket image: Photograph of Saturn and its moons, including Titan.
NASA/JPL/Space Science Institute.
Jacket design by Kelly Blair

Printed in the United States of America
First Edition
2 4 6 8 9 7 5 3 1

To Charles F. Penniman,
my gentle guide, whose life has unified science, spirituality,
and compassion.

—CHARLES WOHLFORTH

To all of the fantastic, dedicated people working so hard
on the Cassini mission and all robotic and human missions
at NASA and other space agencies: you inspire the world
with these journeys and incredible scientific discoveries.

—AMANDA R. HENDRIX

CONTENTS

BEYOND EARTH

THE WAY OFF THE EARTH

Someday, people will live on Titan, the largest moon of Saturn. Their energy will come from burning the unlimited supply of fossil fuels on its surface and their oxygen from the water ice that forms much of Titan's mass. The nitrogen atmosphere, thicker than the Earth's, will protect them from space radiation and allow them to live in unpressurized buildings and travel without spacesuits, in very warm clothes with respirators. They will go boating on lakes of liquid methane and fly like birds in the cold, dense atmosphere, with wings on their backs.

This will happen because, at a certain point, it will make sense. Today, the cold, gloomy Titan skies are unappealing and impossibly distant. We do not yet have the technology to put people on Titan. But the technology is coming at the same time the prospects for the Earth are getting worse. In earlier times, human beings struck out for strange and dangerous new places when their homes became intolerable. If humanity doesn't change course on this planet, a new world free from war and climate upheaval could someday draw colonists to Titan in the same way.

The technology required for a space colony is already visible.

The largest barriers are institutional. An indifferent political establishment. A space agency, NASA, with a culture that squelches dissent and that lacks a coherent goal for human spaceflight. News media that have sold the public a false understanding of the real challenges of space exploration. Going to another planet will be difficult and, without breakthroughs, unacceptably dangerous.

But the ingredients for a space colony are coming together. Experience building space vehicles has spread to many countries and private industry. An Internet-spawned innovation culture that knows how to make new things fast has turned its attention to space. The concepts needed to get us there have been thought out already.

When the moment comes, it won't be the first time human beings have embarked on a voyage that seemed impossibly difficult, expensive, and technically challenging. Our kind repeatedly built new societies in places so remote as to forbid return. When we do it again, we'll probably have reasons similar to those they had then.

As authors, we have investigated science and technology as well as culture and the environment to construct our scenario about space colonization. We have pondered the fundamental issues facing humanity: our response to technology; our will to explore, expand, and consume; and how we treat one another and treat the world we already have. The most important ingredient for space colonization is the human animal: our cellular response to cosmic radiation, our psychological ability to travel for years through nothingness, and our ecological fitness for a new landscape where no organism has lived before (at least no organism we know of). What are we? How far can we go?

Scientists we interviewed often asked if we were writing science fiction or journalism. We never intended to write a work of imagination, but a skeptic would never have predicted what has already happened. We visited a rocket factory floor where private space industry workers were sewing astronaut suits that Captain James T. Kirk would have been proud to wear. Our scenario is not based on a love of cool inventions and inspiring visions. It relies on our knowledge of people's tendency for dumb decisions, selfish drives, and messy politics. Recognizing these predictable truths makes it easier to see how technology could unfold, and more interesting and funnier, too.

We've had tremendous fun thinking and arguing about how this will happen. The work developed while we laughed together for many hours on Skype, Amanda at her office or kitchen table in Los Angeles and Boulder and Charles in a home office facing his snow-buried boat in Anchorage or in the Alaska wilderness.

Amanda works with space technology every day. She has practiced to become an astronaut and has managed equipment to capture the scenery of a world on the other side of the solar system. She has also navigated the bureaucracy of big science, a universe of meetings, travel, and egos like any modern organization. Laboring over the myriad details of new ideas, she has helped make the miracle of space exploration real.

Charles spends summers off the grid on an Alaskan beach and in winter he cross-country skis almost every day. His books seeking to understand the fate of the planet have taken him to the Arctic pack ice with Eskimo whalers and to a Cambridge, England, aviary with supersmart birds.

We're complementary opposites. Amanda brings the science and the wonder, but also an awareness of how technology unsteadily unfolds. Charles brings the skepticism of one who has studied human tragedy on Earth, but also the optimism of one who loves the nature in us all. Amanda would eagerly accept a one-way ticket off the Earth to fulfill her drive for adventure and her vision for the future. Charles can hardly sit still through a red-eye flight and cannot imagine saying good-bye to the snow, sea, and fresh air of this world.

We'll never be rich enough to send everyone off to another celestial body, but it isn't hard to imagine a day when governments or the very rich begin thinking of a spaceship as a lifeboat—or an ark. People are already thinking that way. In 2008, the Svalbard Global Seed Vault opened, deep inside a frozen mountain on an island halfway between Norway and the North Pole, to preserve millions of plant lines in case of disaster or apocalypse. An off-world colony would likewise shield a store of human genes out of the way of any earthly harm.

But, unlike seeds, human beings don't stay the same when you put them in safe storage. An extraterrestrial colony might begin as an annex to Earth, preserving our species, but it would develop into a world of its own, with its own culture, government, and future. Within a generation the Earth could be a foreign place to children born under an orange sky. For them, the smell of recycled air, not fresh breezes, would carry the nostalgic sense of home.

We envision their sky as orange because our scenario for this future leads us to Titan. Why Titan? We have screened each of the places colonists could go to find one where the requirements of human safety and sustenance can be met without direct support from Earth and without end. The process of constructing a scenario led us to this wet, energy-rich object in the outer solar system.

We aren't exactly predicting that a colony will be built there, and we certainly don't know when it will happen. A scenario is a way of organizing an investigation of the future, not a prediction. This powerful exercise generates a thought experiment that anyone can run in his or her own mind, using the hard information we provide. As you come with us through this journey into a possible future, you will find all the facts for your own thought experiment to see if your reasoning brings you to Titan, too.

The book's structure reflects this interplay of hard science and intriguing projections. Alternating sections cover reality and the future scenario. Sections labeled "Present" report the technology and ideas that already exist and tell the stories of the real people bringing space closer. Sections labeled "Future" project a scenario that responds to forces and opportunities that seem to us likely, as well as some whimsical predictions. The book interweaves these two modes to create an integrated picture of what's known and where that knowledge could lead. Readers are free to reach their own conclusions.

Building a self-sustaining space colony is several decades and technology steps away. But many space scientists and engineers think about it, because it represents the kind of adventure that brought them into the field in the first place. And because it raises fascinating questions about today's technology, research, and space industry. Indeed, the goal of moving people to a new planet is the strongest justification for a manned U.S. space program.

We've started with the current state of space science, then asked how we could get to another celestial body and build a colony there, where we would go, and why it might really happen. To be credible, the scenario must respond to these three questions—how, where, and why—with answers that are based in today's reality. That's why the book uses alternating sections: so the present can inform the future for each of the three questions.

The question of *how* addresses the technology of both advanced propulsion and the design of conventional spacecraft as they become commonplace. As the business of spaceflight becomes assimilated into our everyday lives, economics holds the key. The commercial spaceflight industry is transforming our sense of possibility. Using Silicon Valley's money and innovative confidence, it will soon bring mass space products to the market. This industry looks something like the computer business did when Steve Jobs and Bill Gates were leaving their garages: ready for a rapid spread beyond the confines of expensive, centralized government projects and into making space travel safe, repeatable, and affordable. As that happens, the cost of lifting materials to weightlessness will fall dramatically, transforming the practicality of every space endeavor.

The question of *where* brings out the many insights of planetary science, as well as space health, psychology, reproductive issues, and the interlocking needs of energy, shelter from radiation, and adaptation to low-gravity life. Colonists need a place they can survive and support themselves, indefinitely.

This "where" question has two steps: asking what human society needs for survival and picking the best spot in the solar system to meet those needs. Titan's hydrocarbon hills and lakes can provide unlimited fuel. Water and its constituents of hydrogen and oxygen make up half of Titan's mass. With water and energy we can make food, process materials, and power cities.

That leaves the question of *why*, which asks for plausible reasons to leave the Earth permanently.

This isn't a Lewis and Clark journey of discovery, probing the unknown and returning home to fame and glory. Space colonists are more like the anonymous pioneers who headed west in covered wagons to build homesteads. The trip won't be fun and they aren't coming back. Life will be hard and dangerous for at least the first generation. Some exceptional people might be willing to go merely to be the first, but creating a colony will require more than adventurers. A colony will need people, a lot of them, with specialized abilities and the resolve to settle and build. Most of all, it will need a sponsor willing to pay the enormous cost of launching the endeavor. In some ways, the "why" question is the biggest of all.

Often, the colonies of the past justified their cost either by sending something of value back to the mother country or by giving the colonists themselves a way to get away from something bad back home. As for the first reason—making money—the business case for living in space remains murky. Space mining could produce materials that don't exist on Earth or are extremely rare here. Heat-shocked graphite asteroids laced with diamonds. Moon deposits of a helium isotope, He-3, implanted by the solar wind, that could power fusion reactors.

But fusion reactors don't exist yet and interplanetary payloads currently are worth more than diamonds (and we probably don't need that many diamonds anyway). Indeed, no resource we know of justifies the cost of a mission, let alone a colony. The business proposition for colonization could improve with cheaper space travel, a big find by space prospectors, or a new technology that requires materials not found on Earth. For us as authors, these possibilities remain hazy, so in our scenario we have chosen another motivation to drive space colonization: the need to get away from the Earth.

Human beings can't go west anymore. Our planet is full. Our personality as a species suggests some of us won't put up with that situation indefinitely. True, the human family is mixed. Some stay home, fix what doesn't work, or learn to live with nearly intolerable political or environmental conditions. But others break away to make new places, never planning to return. We spread. We have done so since leaving Africa to populate Europe, Asia, and the New World, and since then we have repeated the process over and over again.

The trends in environmental and political conditions on Earth are as important a part of a scenario for space colonization as the technology story, and an important part of understanding ourselves. Along with learning how our bodies respond to space, we need to predict how our societies will respond to a deteriorating environment, increasing political and religious conflict, and widening wealth disparities.

The inspiring and discouraging qualities that make people such interesting animals are already mixing to create the movement toward space. Billionaires are building spaceships to go farther than any man or woman has gone before, at the same time as they make money selling tickets for joyrides beyond the atmosphere.

We've met brilliant young engineers at commercial space companies who are driving down the cost of access to space. They are putting in long hours to make spaceflight a part of everyday life and thinking several steps beyond. They don't work so hard just for a paycheck. These young aerospace workers speak the language of *Star Trek*, attacking huge technical challenges with the dexterity and attitude of hackers. They're unfamiliar with failure and utterly sure they are on the way to space.

Spending time with these people, it's not hard to imagine a day when a huge spaceship will prepare for departure from the end of a long, retractable corridor connected to a commercial space station orbiting the Earth. The scenario may sound like science fiction. But doesn't the future often sound like science fiction—until it suddenly arrives?

1

HOW TO PREDICT THE FUTURE

Predictions about the end of the world have been around even longer than dreams of colonies in space. Sometimes they're combined. When the TV show *Space: 1999* came out in the mid-1970s, people didn't laugh at the premise that by 1999 people would live in a colony on the Moon that would be flung into deep space. Twenty-five years later, when the predicted date rolled around on September 13, 1999, fans celebrated at a convention in L.A. by re-enacting the disaster that had never happened.

Here we are doing it again. We predict that life on Earth will get bad enough and technology will get good enough that a space colony will be built. We're not in great company.

The optimism of the 1950s had people in flying cars and space hotels by the 1970s. The pessimism of the 1970s had us all dead from innumerable causes by 2000. Pessimists and optimists have equally bad records. We never got flying cars, but we did get video phones. We haven't run out of food, water, or energy (yet), but the climate did get warmer and the weather more violent and erratic. Power from nuclear fusion remained perpetually fifty years in the future, but robots are everywhere and they haven't turned against us yet (unless you're a cat and you live with a Roomba).

We're confident of a lot of predictions about things that don't change. Setting an alarm clock is based on a prediction that the Sun will rise. Changeable things can be predictable, too, with the certainty that they will not stay the same. Technological development and human imagination are as likely to continue as conflict and greed. The future is difficult to predict, but it is easy to see that people will be powerful and will use their power, that they will consume resources and will strive for the next idea that brings greatness.

The trick is to analyze that knowledge, teasing out what makes some predictions succeed and others fail.

In 1955, TWA, the airline owned by Howard Hughes, stalled over its decision whether to order the new jet aircraft that would revolutionize travel three years later. "TWA President Ralph Damon is on record as predicting that the jet age is not as close as the jet purchasers like to think," the *New York Times* wrote at the time. "There are only a handful of airports in the world with large enough or strong enough runways to accommodate the huge 100-passenger Boeing and Douglas planes . . . Millions will have to be spent improving runways. Will many cities be willing to make such investments?"

TWA had been around for thirty years at that time, having begun in the 1920s (as Western Air Express) with eight-hour flights from Salt Lake City to Los Angeles. It celebrated its anniversary in 1955 with a contest asking passengers to predict what commercial aviation would be like in another thirty years.

Passengers found contest entry forms in seat-back pockets of TWA's prop-driven airliners, offering a fifty-thousand-dollar prize payable in 1985. The passengers' predictions for 1985 included aircraft that could go 40,000 kilometers per hour (25,000 miles per hour), rocket-propelled hotels, crash-proof planes, flying taxis, helicopter house trailers, and the prediction that there would be no aviation at all, because the world would be inhabited only by monkeys. Or we wouldn't need airplanes because we would live on other planets without much gravity and could fly without them.

Thirty years later, in 1985, TWA found the entries in a vault in Kansas City, Missouri, sorted out the whacky predictions, and gave finalists to a trio of judges, including astronaut Pete Conrad, who had stood on the Moon in 1969. From before the jet age, the contest winner had described the current state of 1980s aviation spot-on, including how aircraft would be used.

"Commercial aircraft," she wrote, "will have ranges of about 5,000 miles and will cruise at about 700 miles per hour. They will be powered by bypass jets since atomic energy probably will not be available to commercial aviation in 1985 . . . Passenger ships will carry about 300 people on long trips and cargo ships will haul about 200,000 pounds' payload."

The winner was Helen L. Thomas, whom the airline found living at the same address in Cambridge, Massachusetts, that she had listed on her entry form. At eighty years old, she had retired from editing research publications at MIT and had no memory of entering a contest. It took airline officials some time to convince her they were serious and give her the fifty-thousand-dollar check.

What did Helen L. Thomas have that TWA president Ralph Damon and all the other contestants lacked when they looked into the future?

First of all, Thomas had no stake in the outcome. She could look at the state of society and technology without an emotional investment. Also, she must have been very intelligent and determined, as the first female in the United States to earn a PhD in her field, the history of science. And that study would have given her a powerful sense of how discoveries develop over time. She had written and researched on aeronautics, too, so she knew about the fan-jet, then on the drawing board, and that it would probably go into use.

Practically, she won simply by looking at existing technology, projecting reasonable improvements, and adding innovations that were on the verge of reality. And she kept in mind economics—the critical human factor of what people need and are willing to pay for. She predicted, for example, that new airports would be built distant from cities, and that executives would use helicopters to get quickly from their planes to their meetings. "The patient who has flown for emergency treatment may land on the roof of the hospital," she wrote.

A *New York Times* editorial in 1986 complimented Thomas's prescience while making fun of a contestant who had predicted we would someday have tourists in space. That prediction was thirty years early, as we've already had space tourists, and many more are coming soon.

The *Times* also highlighted the importance of free enterprise to technological progress. Thanks to airline deregulation, in the thirty years from 1955 to 1986 the cost of a coast-to-coast plane ticket had barely risen, from $99 to $129. Another thirty years later, as this is written, Travelocity has the same one-way ticket for $135. Counting inflation, the cost has fallen 85 percent since 1955.

During the same thirty-year period of aviation improvements, telephones had hardly changed at all. Many homes still had a rotary dial phone, invented in the 1890s. But telephone deregulation had just arrived in 1986, prompting the *Times* to suggest in its editorial, "Let's have a contest to predict the quality of phone service in 2016."

In nearly thirty years since, phones have changed a lot more than airplanes. No one but science fiction writers predicted the computing power and information gathering ability we hold in our pockets now. The market met opportunity and bright people and poured enormous resources into innovation, and we got to enter a new world. Technological possibility was the threshold: the power that pushed us through the door was the desirability of what the technology promised and the willingness to pay for it.

There's a pattern to the timing of changes in technological fields. Big revolutions like the invention of the telephone or the jet happen rapidly, followed by long periods of incremental improvement, and then another revolution. The planes we fly now look a lot like the first successful passenger jets. They're bigger, quieter, and more efficient but have much more in common with the planes of 1960 than the planes of 1960 did with the planes of 1955. Similarly, telephones got better slowly for almost a century before they exploded into a whole new thing.

Maybe it's not a coincidence that thirty years is about the time it takes for a new generation to grow up with technology, take on the jobs capable of disrupting it, and create something totally new. Young people renew the world because they haven't learned yet what's not possible. The clock is ticking for the millennials to show us the way back into space.

The space age achieved its epic triumph in twelve years, from

1957's Sputnik I, an orbiting ball containing a radio, to 1969's Apollo 11, which landed men on the Moon. Then came forty-seven years when human spaceflight didn't change much and astronauts didn't venture beyond low Earth orbit. Here was the pattern seen in commercial aviation and telephones. But where is the next revolution?

A Saturn V rocket like the ones that carried astronauts to the Moon lies on its side in a metal building outside the gate of the Johnson Space Center in Houston. Restorers painted it and put it indoors after years of weathering made it look shabby. It looms massively for inspection like a relic from an age of giants, appearing as enormous and impossible as Roman ruins must have looked to tribesmen of the Middle Ages. The rocket stood thirty-six stories tall, bigger than the Statue of Liberty, and it flew straight up—it is overwhelming to imagine. But even more amazing, it took NASA only five years to bring the Saturn V from conception to successful flight. Even if it had the money, NASA doesn't know how to innovate that rapidly anymore.

After more than forty years in low Earth orbit with little change in human spaceflight, the other planets and stars seem impossibly distant, much less reachable than they appeared to the more optimistic but less technically capable engineers of the 1950s. But habits of seeing shouldn't blind us. The gathering light of a technological revolution can be discerned before it dawns. Helen L. Thomas did it in 1955. Bunches of now-famous billionaires did it as the Internet developed.

Each of us has an internal plausibility meter that we check before deciding what to believe. NASA may have lost its ability to surprise us, but that fact alone shouldn't set the plausibility meter too high. That would be like the mistake made in 1955 by TWA president Ralph Damon when he said jets wouldn't catch on.

The evidence suggests that by being open-minded, smart, well informed, and unbiased we can predict the future of spaceflight, exploration, and colonization.

George T. Whitesides, whose job is to build a spaceline and make money carrying passengers and cargo, believes today's residents of Earth relate to future space colonists as the first Asians building reed boats relate to today's Pacific Islanders. We know we're going somewhere, but the destination and route are difficult to predict. "Assume you are like an African tribesman, and you're trying to imagine how quickly humanity will spread across the globe, with that set of technologies, two hundred thousand years ago."

The assessment makes him sound like a pessimist, until you learn George bought a pair of $200,000 tickets to fly into space from the company he is now running, only months after the first experimental prototype, SpaceShipOne, scored a technology prize by breaking above the atmosphere in 2004. George is both an idealist and a business-based realist. An Ivy League Brahmin whose father is a famous Harvard chemist and inventor, he stands at a unique intersection between the engineers and the enthusiasts.

His boss is an enthusiast. British billionaire Richard Branson wanted to own a space tourism company before Burt Rutan launched SpaceShipOne. He had registered the name Virgin Galactic (like his Virgin airlines and record company) long before there was any craft likely to be able to carry paying passengers, spending years investigating backyard inventors' improbable ideas looking for something to buy. Rutan had been funded by Microsoft billionaire Paul Allen, who also contributed tens of millions toward picking up radio signals from extraterrestrial civilizations through the SETI Institute.

The need for an inexpensive, reusable space vehicle has been the obvious next step for forty years, ever since the conception of the space shuttle. An inexpensive shuttle into space would allow large structures to be built there, including massive vehicles that could go to other planets, and cost-effective space businesses of unimagined kinds. NASA sold the space shuttle to Congress in 1972 with claims that it would fly up to fifty times a year, launching payloads for under $100 a pound, and predicted a 10 percent return on investment. When the program ended in 2011, NASA had spent $192 billion on 133 successful flights with a payload cost per pound (if every shuttle had been at capacity) of about $30,000.

An alternate route came into view as Earth grew an abundance of baby-boom billionaires like Allen and Branson, raised on the space optimism and science fiction of the 1960s and '70s. The $10 million Ansari XPRIZE, which drove Rutan, was funded by an Iranian-American telecom entrepreneur who later also bought her own seat on a Russian rocket to the International Space Station for some $20 million. The rules of the prize called for the first privately funded craft able to carry three people, including the pilot, that could go 100 kilometers (62 miles) high, just above the atmosphere, and repeat the feat within two weeks.

The fragile little SpaceShipOne, the size of a pickup truck, rode to its launching point at 14 kilometers (46,000 feet) under a carrier plane called WhiteKnightOne. It hardly looked like a competitor to the enormous space shuttle and it did not fly into orbit. But the entire project had only cost around $27 million. NASA pays the Russians $70 million a seat for every astronaut its rockets take to the space station. After pocketing the prize and selling the technology to Branson, Allen actually made money on the deal. The economics suddenly seemed real.

Branson immediately began selling seats, predicting he would take passengers up in a larger SpaceShipTwo by 2007. He has made fresh predictions many times since. His company, Virgin Galactic, sold seven hundred tickets, more than $140 million worth, to movie stars and other rich folk. But the company has spent far more. Tickets now go for $250,000 instead of $200,000, purportedly because of the inflation in the decade since they originally went on sale.

Whitesides joined Virgin Galactic as CEO in 2010, after less than two years as chief of staff to the administrator of NASA, a position he had attained through involvement in the 2008 Obama campaign and transition team. He's the sort to instill confidence after years of delays and setbacks, including a fuel test explosion that killed three employees of Rutan's company, Scaled Composites, in 2007 and a test-flight crash in 2014 that killed a Scaled Composites pilot and destroyed the spaceplane. He's precise, with the air of a person who never says the wrong word, but relaxed and able to talk about big issues. He brings a corporate crispness to a Kitty Hawk operation.

The Virgin Galactic offices occupy sun-faded metal buildings along a runway in the desert 160 kilometers (100 miles) northeast of

Los Angeles. This World War II strip next to the desiccated gold-mining town of Mojave was where Rutan found cheap hangar space decades ago and began inventing airplanes, including the Voyager, which flew around the world in 1986. On the table-flat Antelope Valley the town pops up beside the highway like other desert towns, seemingly at random, with few people in evidence and tired buildings begging for mercy under the unremitting sun. Everything stands out as if placed on display in such a barren landscape. Here, that includes a fleet of retired and mothballed passenger jets, a forest of windmills at a huge alternative energy farm, and a menagerie of experimental aircraft set aside like discarded toys.

Starting with Rutan's Scaled Composites, the Mojave Air and Space Port became the nation's center of private space innovation. Engineers for many companies work in hangars around the runway—some of which have vertical doors for wheeling out rockets—with access to clear skies, open spaces, restricted airspace, and privacy. Techs with greasy hands meet for lunch at the old-fashioned air-port diner amid the aerospace memorabilia. There's nowhere else to gather and not much else to do but work.

But if you like building airplanes, like Rebecca Colby, a young engineer we met in a break room at Virgin Galactic, this is the best possible place to be. She spent lots of time making things fly at MIT and now, for recreation, she was learning to fly herself. The day we talked she had spent the morning computer-modeling the strength of a design change in the WhiteKnightTwo carrier plane that will lift Virgin Galactic's spaceship to 15 kilometers (50,000 feet) for launch. She had also been modeling the trajectory of the SpaceShipTwo vehicle itself to determine the best angle to launch it from the carrier.

This was Colby's first job out of college. While she knew it was cool to be making design changes to a spacecraft that had tickets sold to Leonardo DiCaprio, Angelina Jolie, and other famous people, she didn't have anything to compare that to. No one did. Everything about the job was new.

"It's not just that I don't know what I'm doing because I'm young, it's that I might be doing something where there's not an accepted way to do it," she said.

The commercial space industry's culture of newness, excitement, and getting the job done helped George Whitesides believe in what he was attempting. At NASA he had been atop an organization with more than fifteen thousand employees working on many projects but without a clear sense of where they were going or why. Here he had a team counted in the hundreds, but everyone was committed to getting SpaceShipTwo into space.

George doesn't think the Apollo approach will work anymore—whether the mission is Mars or something else—with enormous money poured into a single goal for many years. Political leaders don't have the risk tolerance or perseverance. It is too expensive to make a complex machine safe by contemporary standards for a single use. And when a huge project ends, so does progress. Large, centralized projects don't create change as effectively as many small, rapid advances distributed among numerous competitors and collaborators free to pursue their own best ideas.

"I can make a decision about a technical aspect, and that's that," George said. "People understand the goal and how to get there. NASA is a much different creature. It is an organization that has a funding stream forever into the future."

Whitesides casually walks through the hangar where WhiteKnightTwo and SpaceShipTwo reside, a garage like any other working industrial shop, but for the bright white airplane with gossamer wings that stretch 43 meters (141 feet) wide, longer than a Boeing 737's. The WhiteKnight is a remarkable high-performance aircraft. It has two fuselages, like a catamaran, and four large jets and yet is made of superlight composite material. George said it can climb to 15 kilometers (50,000 feet) without running the engines above idling speed. The WhiteKnightTwo can also bank a turn at 6 g's or go parabolic—flying a big arc in the sky that creates temporary weightlessness in the cabin.

The power, strength, and light weight are needed to haul the 13,600 kilogram (30,000 pound) SpaceShipTwo to its launching altitude while it hangs between the two fuselages. The craft is a six-passenger, two-pilot spaceplane, 18 meters (60 feet) long. The diameter is similar to an executive jet, but without the floor. Instead, the entire interior of the cylinder is given over to allowing the passengers to play during a few minutes of weightlessness on the flight. Rutan's big innovation was the shape and flight surfaces of the spaceplane, which allow it to re-enter the atmosphere like a badminton shuttlecock, without risk of overheating. An oversized tail fin and wing assembly feathers into a position perpendicular to flight to create stability and break speed. Back in the atmosphere, at about 21 kilometers (70,000 feet), the tail section, called the feather, rotates back into a wing for the glide back to the landing strip.

SpaceShipTwo is designed to make it into space at 4,000 kilometers per hour (2,500 miles per hour) but not to go into orbit. Its flight is a quick arc up above the sky and home to Earth again in one and a half hours.

Powered flights with test pilots began in early 2014, but big changes remained to be made. Virgin Galactic changed the fuel for the rocket on SpaceShipTwo in May 2014. That fall, it missed another predicted inaugural flight with passengers.

Then, on Halloween, the crash happened on a test flight operated by Scaled Composites, a company now owned by Northrop Grumman (Rutan retired in 2011). The copilot pulled a lever to unlock the feather at the wrong time, about fourteen seconds early, while the craft was still accelerating to go supersonic. The feather deployed and the spacecraft tore apart. The pilot was thrown free and parachuted to the ground, injured but alive. The copilot died, his body found in the wreckage.

An investigation by the National Transportation Safety Board found that the crash happened because of human error but faulted Scaled Composites for not including more safeguards against the mistake. Because the crash resulted from a pilot's error and not a design flaw, ticket holders seemed to accept it. Only a few requested refunds.

A few months before the crash, touring the Virgin Galactic hangar, we saw the ship that would be lost. On the far side of the hangar from the finished aircraft and spaceplane, a couple of guys in T-shirts were building another copy of the spaceship while listening to the U2 album *Joshua Tree*. Sheets of fresh composite material lay across metal forms to become new pieces of wing or fuselage.

Virgin Galactic had already taken over the construction and ownership of the craft from Scaled Composites, with plans to build five copies. Whitesides said the company will be limited by capacity, not demand, once SpaceShipTwo is flying.

Predictions of how long this would take were wrong, but the prediction still seems accurate that Virgin Galactic will fly and carry paying passengers. Whitesides's market research says millions of earthlings can afford a flight. Virgin Galactic can make money, so long as flying its spaceship remains cool and is perceived as safe. The business is entertainment. No one needs to go up in this craft.

George has spent most of his career in space marketing and promotion, not science. He helped get Zero Gravity Corporation off the ground, a company that for a decade has taken passengers on parabolic flights for sequences of 30-second weightless episodes in the padded, windowless cabin of a specially adapted Boeing 727 (after twelve arcs, passengers are weightless longer in total than they will be on SpaceShipTwo). It is *fun*. Amanda made a series of rides on NASA's version of the vomit comet for research and found it was better than water skiing, sky diving, or scuba diving. A trip on the commercial vomit comet costs five thousand dollars per person, but plenty of wedding parties have done it. In 2014, *Sports Illustrated* swimsuit model Kate Upton floated around the plane, with her tiny bikini top getting an antigravity assist.

The space tourism business looks like a carnival at times. As head of the National Space Society, George flirted with its looney fringes, putting on an annual conference that draws some serious scientists and a lot of eccentrics (Amanda snapped a photo of an attendee in a vest emblazoned with the declaration "Luna City or Bust"). Virgin Galactic's website tempts ticket buyers with the potential of hanging out with Branson on his private Caribbean island. That's what the business demands. Shooting movie stars into space is probably a smart way to make money at this, and floating weightless brides and bikini models get attention. George himself said he planned to honeymoon in space with his wife, Loretta, when he bought their Virgin Galactic tickets in 2004 (she is still at Zero Gravity, and still his wife).

But the glitz is a means to an end. Using its start in space tourism, Virgin Galactic is developing LauncherOne to put satellites in orbit. The step after that could be orbital vehicles for passengers, or building a craft to fly passengers and cargo around the world at extraordinary speeds via space.

"Point to point seems like, by far, the biggest ultimate market for space transportation technology, and so therefore you would think it provides the biggest economic driver for driving the cost per flight down," George said. "That's why I joined Virgin Galactic. Because I think it could be the first big step on this huge route out."

Whitesides is thinking about the market for making any city on Earth a day trip from any other. For those who could afford the ride, the planet would become suddenly much smaller. Such a huge market would bring sustainable new flows of money for taking technology to the next step after that. If he is right, the moment when passengers ride to space with Virgin Galactic could be like the day the first computers left labs and found their way onto ordinary people's desks. At first they seemed like gimmicks or toys, and people talked about using them to store recipes. But once they became a product we used every day, their power and price improved exponentially.

George says the limits are not physical, they are technical and economic. The cost per passenger depends, for the most part, on the cost of the vehicle divided by the number of seats and number of flights, not the fuel or the pilot's salary. George estimates that with the leap from NASA's space shuttle to Elon Musk's SpaceX Dragon spacecraft, the cost of getting a human being into orbit will go down by an order of magnitude, from more than $100 million per passenger to around $10 million. Virgin Galactic, going suborbital, expects to remove two orders of magnitude from the cost, until it makes money at its passenger fare of $250,000. In another decade, Whitesides expects spaceflight to get down to $10,000 or less.

When it costs $10,000 to put a person in space, many other things will rapidly follow.

Often, space scientists avoid predicting beyond the next step of technology. Bringing up the idea of sending people to live permanently off the Earth can be embarrassing in professional circles. NASA scientists skirt far-out ideas espoused by amateur space enthusiasts, even when the nonscientists talk about exciting topics. The word "colony" itself is frowned upon (it's OK to say "habitat" instead). The caution can be understandable. Amateur space gatherings tend to look more like sci-fi fan conventions than reputable scientific meetings, mixing reality-based talks with sessions on psychic travel and such. Although the media make little distinction between fanciful ideas and real projects, planetary scientists protect their reputations by steering clear.

Some engineers and scientists manage the risk to how they might look using the same strategy employed by preteens in middle school: they keep their imaginations under wraps. Consequently, the scientific literature on topics like space communities and artificial gravity has been largely dormant since the 1970s. Amateurs cite those few reputable books and papers from the past the way fan fiction writers draw on the official *Star Wars* movies.

But NASA needs a goal. Focusing on the idea of putting a colony on another planet could improve nearer-term missions. (There, we dared use the word "colony"!) NASA's history of thinking only of its current mission has repeatedly left it with a full-stop loss of momentum after each success, owning single-purpose equipment irrelevant to the next project. Even an imaginary space colony decades off would provide a goal, a lodestar, to help align mission planners' designs to the future. Barely managing a quick visit to a Mars habitat wouldn't do that. Besides, taking a long view would be the best way to meet the challenges of eventual space colonization. Problems could be picked off one at a time on missions to Mars, the Moon, and even the International Space Station. The guiding principle would be to build capacity for long-distance, long-term space travel and habitation.

But why build a space colony at all? It might never be needed. Exploration, yes. Even temporary bases on other planets. But why send people to live somewhere else? Earth is a miracle. People can live here without any technology at all, at least in some places. In sunny spots where fruit and fish are plentiful, human beings never have to leave Eden. No place on Earth is as bad as the best place on any other planet, moon, or asteroid in the solar system. Even the Antarctic offers a breathable atmosphere, shelter from harmful radiation from the Sun and the cosmos, and a gravitational field to which human bodies are properly adapted.

But people haven't always cowered safely in the best places.

For early indigenous settlers, relying on food they could harvest and building their own shelter as they traveled, the frigid Arctic was as hostile a place as Mars is to us now. We can stay put on Earth, breathing its free air, but the people of the past also could have stayed in small tribal groups in Africa, harvesting fruit from the trees. They didn't. No one today lives without technology, and hardly anyone ever has. Ancient bones usually show up with stones that were used as tools or weapons, the technology of the time. As early humans figured out new ways of doing things they began to migrate from Africa into new lands that their technology made accessible and habitable.

When our early ancestors invented boats and navigation, they set off for unknown islands in the South Pacific, crossing spaces as wide and unknown as the solar system is to us. They voyaged over immense spans of ocean toward tiny islands they might not find and couldn't be sure even existed. They explored blindly, unable to communicate with loved ones, in small boats vulnerable to storms and with a limited time afloat. Somehow, some survived, found land, made families, and built new societies that thrived for thousands of years. Would any astronaut's mission be as daring or uncertain as the first Pacific people's search for new islands?

In the north, the technology of sewing warm clothes and boots allowed people to occupy the bitterly cold Arctic. They trekked over moving sea ice and frozen tundra, dependent on hunting for all their needs, and built homes in the most hostile environments on Earth, homes they would heat with seal oil warmly enough so they could go without clothing when indoors—like astronauts who wear spacesuits during extravehicular activities and then strip down to civvies when they come inside the space station. The Inuit dug their houses out of frozen ground with handmade tools, supporting their sod roofs with rib bones from 18,000-kilogram (40,000-pound) whales they killed with spears and pulled out of the ocean using walrus-hide ropes. Would any space habitat cost modern society as much, comparatively, as a village would have cost the Inuit?

If we spread beyond Earth, it would probably be for similar reasons as our ancestors had for spreading across this planet. In the Arctic, archeological evidence shows how a worsening climate led to technological innovations that allowed a group known as the Thule people to wipe out earlier cultures and spread eastward from Siberia to Greenland in a few generations. Rapidly changing climate conditions and population growth probably drove much of humanity's rise and spread over the last quarter million years.

We haven't conquered the climate. Weather still has the ability to make humanity puny. But human carbon emissions have set the stage for climate change, scarcity, and conflict bigger than any that drove the earlier migrations of our species.

It would be far easier and less expensive to protect the Earth than to move people to another planet. But no one on Earth has the power to make that decision. Individual nations and even wealthy people could advance the future of space travel to move outward, but to halt carbon emissions would require cooperation by the entire species.

If we assume that humanity doesn't change its path, what might happen? That question has guided our exercise in drawing a scenario of space colonization, even at the risk of asking embarrassing questions that many scientists avoid. We would certainly save this planet first, given the choice. But, in case humanity does not, we've dared consider the next human migration.

Through the chapters of the book, a future scenario traces our ideas about how that difficult path might develop. We won't nail all the details or the timing. Unexpected events always influence the flow of history. But we have connected a chain of predictions that reasonably fits the facts and that we believe will be persuasive to other open-minded people.

That scenario begins here.

FUTURE

The increasing reusability of spacecraft followed the trend of computers, which shrank from huge mainframes only a government could afford to cheap, powerful tools available to everyone. And, in the same way that computers changed our lives, passenger spacecraft opened up vast possibilities. And money flowed in.

The first small passenger spacecraft were a hazardous lark, but when they were flying regularly, investment flooded into the field like a dot-com boom. With money from Wall Street to burn, ultra-capitalized companies raced to build suborbital spaceplanes to connect any two spots on Earth, airport to airport, in a few hours or less, or coast to coast in ninety minutes. Airlines and new spacelines ordered the planes, leaving behind the quaint idea of a company such as Virgin Galactic constructing and flying its own.

Flights began as superluxuries, symbols of wealth and power like today's private jets. Rappers bragged of weightless parties. Overpaid executives drove the market. Based on the value of their time, it paid to use a plane that could turn all meetings into one-day affairs, worldwide. Once some corporate titans were doing it, others had to follow suit to maintain their prestige. More flights and competition brought the cost down until it became accessible even to upper-middle-class vacationers.

As incomes continued to stratify, wealth concentrating at the top, an entire social class no longer considered taking a conventional jet on an overseas or coast-to-coast trip. Once you had enjoyed the speed and convenience of a spaceplane, it was unthinkable to spend an exhausting day in a first-class airline seat just to get to another continent. As unthinkable as switching from broadband Internet back to a slow dial-up connection. The spaceplane became the new normal, no longer a luxury. A new baseline of necessity had been set.

Travelers on conventional jets became an underclass, their airport concourses dated, with dirty carpet and dodgy loiterers, like today's bus stations. Businesspeople and well-off vacationers walked through sleek new facilities to get on quick suborbital flights. A New York banker could leave for Shanghai at 4 P.M., have an hour-long meeting there at 7 A.M. and be home by bedtime at 11 the same night he left New York. A couple could escape the rain of London for a romantic weekend at the beach in Australia.

Spaceplane travel usually involved an extended period of weightlessness, which caused nausea for some travelers and required use of an intimidating suction-powered toilet. Travelers concerned about the bathroom experience could gain confidence with an hour-long preflight toileting class and certification card. Flight attendant safety briefings included, along with the other routine warnings, instructions in use of the vacuum-powered puke bags, and a suggestion that everyone go to the bathroom before separation from the carrier plane.

An announcement at separation would warn everyone to secure their eyeglasses, tablets, and any other loose items, but seasoned travelers already wore lanyards tying down their things and ignored the announcements. In-flight food and beverage service were hardly worth the trouble, although flight attendants were supplied with handholds that allowed them to float their way through the cabin to ask passengers if they wanted anything.

The spaceline industry manufactured new billionaires. Investors looked for the next step in spaceflight where they could place their bets. Venture capitalists began examining opportunities for space colonization.

The world economic boom that brought weekend trips beyond the stratosphere had also increased carbon dioxide emissions to warm the climate. Technology greatly reduced the carbon requirement to move cars and airplanes, generate electricity, and manufacture products, but the increased wealth of the world's population meant more people drove and flew, used power, and went shopping—so the sum of emissions kept rising.

The climate had obviously changed, with giant storms and heat waves disrupting travel and business more frequently. Property damage accelerated in Westernized countries beyond the point of easy repair, and some shoreline communities were abandoned when governments wouldn't pay to protect them. Too many storms were coming too fast to keep rescuing property owners. Cities saw a looming fight for life and began constructing seawalls. The rich began building compounds at inland sites, hardened against disasters and defensible against potential political unrest.

In less developed countries, the problems were much worse. News stories of rising seas, drying rivers, parched crops, and destructive storms competed with familiar tales of extremism and civil war. Famines, epidemics, and population migrations began happening too frequently for the suffering to be mitigated by international aid. Humanitarian efforts couldn't hold back the tide. National security planners instead developed a policy of containment. Noting the relationship of climate suffering, political instability, and violence, they studied ways to keep populations in crisis concentrated rather than allowing the contagion of militancy to spread.

For those wealthy people living on Earth in buildings hardened against weather disasters and walled and defended from political unrest, moving off the planet began to seem, while perhaps unnecessary, certainly plausible. Working toward a space colony could be a sensible insurance policy against whatever might happen next.

PRESENT

Accurately predicting the weather more than two weeks in advance is probably impossible, no matter how powerful computers become or how many sensors are placed around the globe. Too much chaos is built into natural systems. The Earth's complex of air, land, and ocean explodes tiny, imperceptible changes over short periods of time into enormous weather differences. And, while theorists might be able to figure out how a butterfly's wing causes a hurricane, they'll never be able to say in advance which one will do it.

Human beings are even more complex than the weather. As much as we think we know about ourselves, and as much time as we spend listening to experts predict what will happen in the economy, or politics, or sports, we're also smart enough to trust these predictions only for likely and near-term events. We create massive, coercive systems to make behavior predictable—to make absolutely certain, for example, that monthly mortgage payments will arrive without fail. Yet millions of people default, and in 2008 the entire economy fell apart because predictions about mortgage repayments were wrong. Even the biggest events are beyond our predictive abilities. Social scientists and politicians were astonished by the fall of the Berlin Wall and the collapse of communism in 1989.

If predictions of weather and human behavior are so uncertain, how do we use them to come up with a scenario for space colonization? For the most part, we've done it by following the example of Helen L. Thomas, who accurately predicted the future of aviation in TWA's 1955 contest.

First, we take current trends and project them continuing. This approach to prediction doesn't always work, and we hope it's wrong when it comes to climate change. We hope the world will come to its senses and embark on a crash program of carbon reduction. But there's little evidence of that happening now, and continuing on the current trend of emissions will soon produce the kind of patchy but extreme events that can seriously disrupt life and the economy— arguably, we've already reached that point.

For space technology, this strategy means that we assume technologies on the horizon will come to fruition. Again, we may be wrong. Maybe commercial space development will slow down due to a disastrous accident or a world economic crisis. But even if the United States were to lose its nerve in space entrepreneurship, other nations more tolerant of risk would eventually pursue the opportunities. Short of a catastrophe that wipes out centers of wealth and technical know-how, it seems likely that new spacecraft will arrive and be less expensive.

Next, like Thomas, we keep it simple. She didn't call for a complex chain of events. Her predictions were based on the expectation that most people would do the things that made the most sense. Occam's razor is the mental rule of first choosing the simplest explanation for a phenomenon. It helps keep scientists and journalists out of trouble.

Smart people can go astray when they forget the rule that the obvious thing is usually what happens. Predictions of climate change are an important example. Burt Rutan, surely a very intelligent man, has declared that the science of human-driven climate change is a conspiracy to allow the government to take more control. But the physical mechanism through which humanity's carbon dioxide emissions warm the atmosphere is simply beyond question. On the other hand, a very complex theory would be needed to explain a counteracting physical system, and, on top of that, how every major scientific society has been induced to hide the truth. The simple explanation makes more sense: that the world's scientists studying the issue have reached the same conclusion because it is best supported by the evidence.

Good predictions also avoid being too specific. Helen L. Thomas didn't predict particular models of airplanes thirty years in the future, she gave general technical parameters that would be achieved. Climate science works the same way. While weather cannot be predicted more than two weeks in advance, climate averages can.

We can predict with confidence that Hawaii will always be warmer than Alaska on average, although there may be days when somewhere in Alaska is warmer than somewhere in Hawaii. Climate scientists look broadly at the forces powering the climate, add known changes, and then predict broad trends of change. This has worked. The global warming predictions of the first, very simple computer climate model have proven accurate almost fifty years later.

In our scenario we've taken some poetic license to illustrate our ideas with fanciful specifics, but we've kept our timeline general. We're not saying exactly when these changes will occur. Nor do we predict the precise events. The details are intended to illustrate broad patterns that we can see developing in the future. The core of our scenario lies in general ideas: that spaceflight will become cheap and available, that life on Earth will become difficult and scary, and that those trends will motivate an aggressive but disorganized movement to colonize another world.

Many people are thinking this way. A 2013 National Academy of Sciences study brought together scientists and intelligence officials to predict how climate change could stress political systems and create conflict. They noted that conflict already is developing around climate-related scarcity—for example, between Pakistan and India, which both have nuclear weapons, and which both rely on the Indus River, now subject to droughts, and which is eventually vulnerable to drying up with the loss of Himalayan glaciers.

A consensus of expert prediction has developed for commercial spaceflight, too. People are investing their money in it. They believe a vigorous, young, competitive industry of very bright people can create something that now sounds like science fiction. It's a phenomenon we've lived through several times in other industries. Capitalists are betting on it happening again for routine spaceflight. The most imaginative are looking beyond that, to the other celestial bodies they want to visit when the change happens.

We've connected these trends. And we've asked, Where will we go?

2

THE INNER SOLAR SYSTEM AND THE PROBLEM WITH NASA

Mark Robinson helped find places on the Moon to provide human shelter. These spots would overcome challenges of radiation and micrometeoroid exposure, water access and temperature range, potentially allowing for use of an inflatable habitat to create a low-cost Moon base. Which, at the beginning of 2010, remained NASA's goal. Except you weren't supposed to call it a base.

"The strategy was to build, first they said 'base,' and then that had military connotations, from a purely PR or political standpoint, so then everybody was ordered to use 'outpost,' because you couldn't use 'colony,' because that suggests little kids running around going to elementary school and stuff like that. And I don't think you were allowed to use the word 'permanent,' or 'semipermanent.'"

The Lunar Reconnaissance Orbiter was launched as part of President George W. Bush's Constellation Program, a 2004 plan to send astronauts to the Moon as a step toward a trip to Mars. Mark liked the plan. A planetary geologist at Arizona State University and leader of the orbiter's camera team, he thought it made sense to develop capabilities and harvest resources on the Moon, three days away, before setting off for a yearlong Mars voyage.

He also was sensitive to using the right words. He had seen the previous mission to the Moon and Mars, announced in 1989 by the previous President George Bush, canceled after years of work and many billions spent. He didn't want that to happen again.

What Robinson and his colleagues had found were steep pits that reminded him of the sheer-walled openings to lava tubes on Earth. On the same floor as his office in Tempe, Arizona, he asked the team at the orbiter camera's Science Operations Center to find opportunities to look at the surface from an angle instead of vertically. In the supersharp images that came back, he could see that the pits had overhangs of rock, providing shelter from directly above, as they would if caves opened off of them. He couldn't see into them far enough to determine if the caves continued horizontally under the surface, but that would make sense.

A ceiling of rock would protect astronauts better than lots of heavy shielding and would keep the temperature steady, in the shade, which would make a habitat easier to engineer. Without the need for shielding, the roof and walls of a habitat could be thin and light, greatly reducing the expense of lifting the shelter off the Earth. The mission might save billions if it could cut so much weight from the launch.

Pits also could be the key to gathering water on the Moon. The Moon's surface is dry. The Sun scorches it to a daytime temperature that vaporizes water quickly. But every day, wet objects collide with the Moon—water-rich meteoroids, comets, and such. Also, the surface is weathered by a constant bombardment of protons from the Sun, which can produce molecules of water from the elements in the lunar rock and dust. Some water probably falls into eternal shadow in the bottom of pits, where the temperature never rises above 40 Kelvin (-388 degrees Fahrenheit), close to absolute zero. There it would accumulate, hard frozen, practically forever, so that even a tiny annual deposit would eventually become a large pool.

Water is useful for more than drinking. It is an ideal material for radiation shielding, and its constituents, hydrogen and oxygen, can be separated and used to make breathable air or to produce energy. Robinson envisioned astronauts living in the pits or caves and emerging from their habitats in spacesuits to mine the water, preparing for a trip to Mars. With the Moon's gravity only a sixth that of the Earth, lifting the water and spaceships from the Moon on the way to Mars would be relatively easy. At the same time, the astronauts would learn how to do productive work on an alien world in ungainly spacesuits.

Whether all this would work isn't certain. There's a lot to know before the plan could be mapped out in detail. But that lack of knowledge is part of the point. Humans haven't been to the Moon since 1972. The Apollo astronauts spent a total of 3.4 days actually working on the lunar surface, over all six missions, and covered 95 kilometers (59 miles) in their rover, all of it around the center of the near side of the Moon. Since then, human beings haven't left low Earth orbit. When he and others found the caves, Robinson was fired up with the thrill of discovery and the excitement of sending more probes and then astronauts to the Moon. He has a plan to send a robotic lander into a pit that would dispatch mini-bots into the caves.

But President Obama canceled the Constellation Program to go back to the Moon. The Lunar Reconnaissance Orbiter would no longer function as the advance guard of manned exploration. It was transferred to the planetary science branch of NASA. Mark still examines the lunar surface, but without the objective of planning for a manned mission.

The pattern of a quarter century had repeated. A pattern of stepping forward, then stepping back, changing course and then reconsidering. If the United States hopes to send human beings beyond the Earth, we must diagnose our inability to choose a goal and stay the course.

On January 28, 1986, the Space Shuttle *Challenger* disintegrated seventy-three seconds after launch. The mission that day had reflected NASA's original conception of the shuttle as a routine form of transportation that would make space cheap and accessible. The shuttle was carrying a communications satellite and a social studies teacher, Christa McAuliffe, along with six professional astronauts. Rushing the launch to fulfill an unrealistic schedule led to the crash.

The moment was indelible for both of us, and for many others. Amanda was in the midst of taking a high school physics final in Pasadena. The class was her favorite, and the only one that seemed relevant to her persistent dream of going to space, but the news made the test almost impossible to finish. Charles was a college senior in New Jersey, in his car, listening to a radio DJ who announced the news at first with disbelief, as if it were a joke. A special kind of shock and sadness fell over the campus, grief combined with disillusionment.

For young people raised on the glory of NASA's early achievements, the news of the *Challenger* disaster hit with the punch of a profound loss of innocence. The revelations that followed made it worse, with evidence of engrained mismanagement and suppression of dissent at NASA. Five engineers, especially Roger Boisjoly, of rocket contractor Morton Thiokol, had warned superiors that the weather at the launch site was too cold and that rubber O-rings in the boosters had not been tested at that day's temperature and would likely fail. The rings were clearly defective, having burned almost through on previous flights. In a conference call the night before the launch, NASA managers intimidated the company into withdrawing its objections. The exact problem Boisjoly warned of caused the explosion. When he later told investigators what had happened, he was ostracized at work and forced from his job.

The *Challenger* disaster happened roughly halfway through the history of NASA, counting from its founding in 1958 to the present. The first half is the stuff of legend. The second half, since 1986, is easy to follow by reading the reports of blue ribbon commissions that repeatedly tried to get the agency back on track. Each commission's report came to be known by its chairman's name: the Rogers report, the Ride report, the Paine report, the 1990 Augustine report, the Aldridge report, the 2009 Augustine report, and so on. The 1990 Augustine report found NASA in a bureaucratic rut, demoralized, and trying to do too much with too little money. It correctly predicted another space shuttle crash. Subsequent reports found NASA in much the same state. When the *Columbia* disintegrated on re-entry in 2003, NASA culture was again cited as a key cause.

The first President Bush, apparently copying President Kennedy, grandly ordered a mission to the Moon and Mars in 1989 but didn't gather political support for the enormous cost. Congress never funded the plan. NASA abandoned the Moon and Mars plan in the 1990s, but in 2004 the second President Bush brought it back, this time as Constellation. Again many billions were spent until the project was canceled in 2010, behind schedule and with no likelihood that enough money would be available to finish it. The Obama administration then adopted the idea of catching a small asteroid in a bag and dragging it into a new orbit around the Moon, called the Asteroid Redirect Mission, or ARM. At NASA, many workers secretly thought that this was a foolish idea that would never be carried out (the design has been improved, but the operation still seems unlikely to actually occur). They work on projects essentially without a sense of an overall mission.

But, as Robinson points out, a lot of fascinating work has happened along the way. The lunar orbiter may not be preparing for a Moon landing, but it is producing great science. Although he feels dragged down by endless teleconferences and tortured by paperwork, Mark was working every day studying the Moon and Mercury with planetary probes flying in the void of space. It's cool and interesting. "You have to be patient. The satisfaction is ten times greater than the frustration."

Mark's path started at an Alaskan gold-mining prospect. He was studying political science and art at the University of the South, working during the summer at a gold camp at Berners Bay, in the coastal rain forest near Juneau. When he received his bachelor's degree he had no job prospects. "My degree makes me really good at cocktail parties," he said.

So he returned to Alaska to work cutting brush, digging holes, and carrying rocks for a geologist seeking gold. In the process he became interested in geology. Pursuing a geology degree at the University of Alaska, Fairbanks, he stumbled upon data from the Mars Viking lander and found he had an aptitude for mathematical image processing. He ended up as a geologist specializing in the surfaces of other planets.

He has plenty to study on the Moon. Questions remain in the field that a five-year-old would ask. Where did the Moon come from? There's a leading theory—that a planet-sized body hit the Earth and the Moon formed out of materials from the collision—but other theories have almost as much evidence to support them. Mark will be happy for a long time trying to tease out the truth from the images, even without having people on the Moon.

But he still dreams of sending humans off the Earth to explore. Many people who work in space science seem to. If it didn't matter, the anger and frustration wouldn't run so deep over NASA's long-term institutional failure. Mark said smart people are leaving NASA, or never applying. The 2009 Augustine report quoted an e-mail that came to the committee chairman: "I am an aerospace engineering master's candidate. [My classmates'] options are working for monolithic bureaucracies where their creativity will be crushed by program cancellations, cost overruns and risk aversion . . . It is no surprise that many of them choose to work in finance."

Robinson gave a presentation to a NASA planning meeting in which he argued that the Moon is still on the critical path to Mars, and that catching an asteroid is not. NASA scientists presented the opposite point of view, extolling the ARM idea.

"During the break one of the key people who had given the other side of the coin came up to me and told me, 'Mark, I'm so glad you said that.' And my response was, 'Why didn't you?' And the answer was, 'Because I can't.' And it's really quite disconcerting that inside NASA there's this attitude that you're not allowed to speak the truth. And if you go back and pull off your shelf the report after the *Columbia* mishap, the investigation report, and after the *Challenger,* both of those reports say that NASA needs to change its culture, and people need to know, if something's not right, they need to be allowed to speak up. And you're still not allowed to speak at NASA, for fear of losing your job."

The *Challenger* disaster produced an object lesson for many other organizations. The movement in the broader society to open and flatten management structures happened over the next two decades. Roger Boisjoly, who had tried to stop the launch, spent the rest of his life teaching ethics as a guest lecturer at engineering schools. But NASA remains a massive, organizationally complex pyramid, struggling to focus, underfunded for its many missions, and increasingly choked by secrecy rules.

At the Johnson Space Center, the feel is purely retro. The architecture is the dreary weathered concrete of 1960s college buildings, a style thankfully gone by now from most campuses. A row of long-disused TV trailers sits outside mission control, their network logos partly worn off by the weather. Museum pieces of past glories are everywhere, including the control room used for the Apollo landings, which has been kept completely intact. For an extra fee, the tourists can walk through it. But evidence is scant of the technology that is changing the world today, with power distributed among individuals using cheap mobile devices and the creativity that power unleashes. It took ages for NASA to allow astronauts on the International Space Station to use iPads.

The space station is NASA's one major accomplishment in human spaceflight since the shuttle. Building it in orbit became the sole purpose of the shuttle. Conceived during President Reagan's administration, the station's purpose also evolved. After the fall of the Soviet Union it became the International Space Station, the ISS, and one of its major goals was to bring nations together in a shared project. When NASA retired the shuttle, only the Russians had the ability to get people to the station. Today, all NASA astronauts take intensive courses in Russian. But well before Russia invaded Ukraine, it was clear that politics would be the dog wagging the space station tail, not the other way around. Operationally, the crews had already segregated, with the Americans and Russians sticking to their own areas at opposite ends of the station and doing their own work with limited interaction.

The 2009 Augustine report, which influenced Obama to cancel Bush's Constellation Program, called for extending the life of the ISS. After twenty-five years of planning and construction and $100 billion in cost, the original plan to crash it into the ocean after only five years of use didn't make much sense.

The purpose of the station had become to learn about human spaceflight to enable long-range exploration. Although the destination remains hazy, that dream is the real, unspoken goal of almost everyone at the Johnson Space Center. Astronauts, engineers, and scientists talk all the time about going to Mars, or about building capabilities so that when a destination is chosen they will be ready for wherever America decides to go in space.

And what will be the purpose of that journey, wherever it goes? One of NASA's fundamental problems is that the reasons given for human space exploration haven't inspired Americans to invest the money needed. In the 1960s, the United States diverted a significant portion of its gross domestic product to go to the Moon. What would motivate us to do that again? The 2009 Augustine report rolls out a familiar list of reasons: to inspire young people, to learn about other planets, to develop technology, and to improve international cooperation. Good things, but is a trillion-dollar mission to Mars the best way to achieve them?

The Augustine report does go on, however, after that routine list, to mention a much bigger final goal. And it does so with a much greater ring of belief and therefore of credibility:

"There was a strong consensus within the Committee that human exploration also should advance us as a civilization towards our ultimate goal: charting a path for human expansion into the solar system. It is too early to know how and when humans will first learn to live on another planet, but we should be guided by that long-term goal."

The first key to a successful colony off Earth would be finding a livable destination. Decades of productive unmanned planetary explorations have provided most of the information we need to screen the choices. This knowledge has grown from the collective curiosity of scientists working with no other purpose than the desire to know. But if we decide to send people off the Earth to live, that pure science will suddenly become highly relevant and, in fact, essential for survival.

Human space exploration to the Moon, Mars, and asteroids is difficult enough. Planners design those missions like camping trips, with a compact package of tools and supplies, irreplaceable after leaving Earth's orbit, that allows survival for a discrete amount of time. Called sortie missions, those trips are designed with return in mind. Colonists would be headed into space to build communities where they would have to forge their own tools.

For humanity to occupy a new home, colonists would have to cut the umbilical cord from Mother Earth. They would need a new, independent ecosystem suitable for life, like Earth, able to provide habitat, energy, and resources.

Filling in that matrix of qualities for a potential colony is an enlightening puzzle.

As to habitat, the environment of the new planet should be survivable with as little technological intervention as possible. This is a matter of degrees. People living anywhere off the Earth would need airtight shelters, but some space environments would be safer and easier to engineer than others.

On one end of the scale, unprotected space would be completely hostile to life, requiring pressurized living quarters able to adapt to temperature extremes, heavily shielded from radiation and equipped with artificial gravity. Planets without an atmosphere could provide gravity and firm ground upon which to build, but the hazards would remain of radiation and sudden loss of breathable air in case of a breach of the shelter by an accident or micrometeorite. Planets with an atmosphere, if thick enough, could also provide shelter from radiation and meteoroids, but the challenges of temperature and atmospheric toxicity would remain.

On the other end of the scale, Earth's tropical and temperate zones are a practical paradise. Human beings can live there without modern technology. Or at least, they can unless climate change, nuclear war, pollution, habitat destruction, and mass extinctions make the Earth more like the other planets. A potential future of extreme temperatures, radioactive contamination, toxic air, and a debilitated biosphere for producing food would present challenges similar to those we would have on a space colony.

Next, energy. It is the basic stuff of our lives and our power over the world. On Earth, we harvest energy primarily from plant photosynthesis. Plants and algae capture photons from the Sun and store the energy chemically—as food and biofuels—and have produced, in ages past, the fossil fuels we burn today.

Plant photosynthesis couldn't power a space colony. It is too inefficient. Too much land and sunlight would be needed (more on that later).

But even if space colonists couldn't rely on conventional photosynthesis for energy, their success would depend, like our own, on abundant energy harvested with locally available materials. Nuclear fuel or solar power cells brought from the Earth could work only as a bridge to colonization. A self-sustaining colony would eventually need to replace those sources with equipment built and powered on-site. A small population on a new planet could not soon replicate the products of mature manufacturing industries on Earth. Simple technology would hold the most promise: wind, tidal, geothermal, or chemical reactions that release energy (burning fuel on Earth is a chemical reaction).

Resources are the basic materials of life. The colonists would need water, nutrients for plants, people, and animals, solids that could be used for building shelter and making tools, and gases that could be made breathable. Starter supplies could be brought from home, but a sustainable colony would eventually have to produce its own. Colonists could leave their settlement to find some elements needed only in small quantities—for example, metals for electronics or nutrients, such as iron, copper, and zinc, could be harvested from metallic asteroids. With plenty of energy, colonists could chemically transform some local resources into what they needed, making water, oxygen, and plastics. But without the right feedstock to work with, the colony wouldn't survive.

Reviewing what we know about the solar system's inner planets, the prospects of checking off many boxes in this matrix of human habitability don't look promising. And that isn't for lack of looking.

The Mars Curiosity rover can see, touch, and smell. In many ways, it is far more capable than a human, with seventeen cameras and a 2-meter (7-foot) mechanical arm that moves a hand with five devices for examining samples, including an alpha particle X-ray spectrometer. But the rover has no judgment. Managing it is like handling a marionette with strings 225 million kilometers (140 million miles) long. And it is so valuable, so special, that deciding which string to pull and how hard requires the input of hundreds of people every day.

Radio signals, which travel at the speed of light, take from four to twenty-four minutes to get to Mars, depending on the varying orbital distance, so it would be impractical to drive the rover in real time. An obstacle might not be visible on a screen on Earth until after the rover had already run into it, and instructions to avoid it would be twice as late. Instead, scientists and engineers driving the rover must send a full day of instructions all at once, let the machine carry out the moves on its own, then absorb the information it sends back to decide the next day's journey. If a hill or other obstruction is in the way, the rover can go only that far before it stops, looks around, and sends the pictures back to Earth for a team meeting in Pasadena to decide what to do the next day.

This is where space exploration is happening, in a nondescript building in the hills north of L.A. Space exploration looks surprisingly like other office work. The team meets in a suite of offices with fabric-covered partitions and talks endlessly on conference calls. Justin Maki showed us around Curiosity's offices at the Jet Propulsion Laboratory (JPL) in Pasadena. He had led the team that designed the cameras, giving him a role among 486 scientists and engineers at institutions around the world who help direct Curiosity. Workers sat at desks and talked on speakerphones, with thirty or forty people on the line at any one time from all over the United States and from England, France, Spain, and Russia.

The day started with a call at 8:50 A.M., Pacific time, to talk about the previous day's data, coming after dinnertime for scientists calling in from Moscow. The call reached its maximum size at 10:15, with discussion of what to do next, when scientists negotiated and horse-traded to bring the focus to their own interests. In two hours, the discussion turned to active plan assessment, making sure the ideas for the next day were practical and would "fit into the bag" of the rover's capabilities and time. Next came the science discussion, conversation with the engineers, and testing the proposed movements on a computer with a three-dimensional model of the Mars landscape. Testing moved the rover like an avatar in a video game, trying out every step of the next day's plan. If needed, a physical mock-up could drive in JPL's Mars yard as a further test. Finally, as the sun set in California, the bundle of instructions, the command sequence, was ready to send to Mars from NASA's Deep Space Network, with antennas spaced around the globe.

The conference call lasts twelve hours, every day, with all the familiar conference-call annoyances—barking dogs, echoes, noisy multitaskers. It used to happen at night, too, as the scientists observed the Martian day, which is thirty-nine minutes longer than the day on Earth. That pushed their sleep cycles around the clock, as if traveling half a time zone each day, so that the team's work schedules coordinated perfectly with their home lives just one day out of thirty-nine. Eventually NASA had mercy and adopted the Earth day for the Curiosity team.

The rover's top speed, including mandatory stops, allows it to cover only 86 meters (282 feet) a day, but that's a lot when you consider how closely it looks. Imagine examining a football field with a magnifying glass while riding a wheelbarrow. When an interesting rock or shadow appears on the horizon, or a stripe of unusually colored sand, scientists argue about whether to visit and investigate it, each pushing for use of an instrument built by his or her own group among the twelve aboard Curiosity.

Everyone hopes to find a fossil or some other unequivocal clue that Mars once hosted life, but the discoveries so far have been about the kind of minerals and weathering that made the rocks look the way they do, sometimes with signs of ancient water. From these tiny pieces, answers may come that say something much bigger. But so far, the pieces increasingly suggest Mars is an unsuitable place from the perspective of organic life, including human beings.

NASA has been studying Mars with spacecraft for fifty years. Mariner 4 saw evidence of past water in 1965 in some of its twenty pictures. The era of the rovers began in the 1990s, inspecting the old streambeds and clay deposits more closely.

At that time, NASA was struggling to recover from the *Challenger* disaster and remake itself as a leaner, more modern organization. As a report from the time said, "For NASA to remain viable and credible, it must become more business-like, treat cost and schedule as important as Mission performance, and deliver on time for the advertised cost."

NASA administrator Dan Goldin launched the Faster, Better,

Cheaper initiative for planetary science in 1992, adopting management ideas others had learned from NASA's earlier failures. Organizations were focusing on areas of strength, avoiding unnecessary complexity, empowering individuals as members of teams, reducing bureaucracy and top-down management, adopting new technology to lower costs, and orienting workers toward goals rather than jobs. Faster, Better, Cheaper tried to copy Silicon Valley, moving NASA from the retro space age to the new-millennium information age. The Internet revolution was happening, enabled by a culture of innovation in which periodic failure was expected and could even be seen as a career plus. Goldin said it was OK to fail. With many more less-expensive missions, taking risks would be rewarded over time even if some unmanned missions didn't make it.

The projects in the first wave of Faster, Better, Cheaper were hugely successful. It's fun to read about them on long-neglected web pages that bring back the clunky look of the Internet in the 1990s. Mark Robinson got involved in some of these fast-turnaround missions early in his career, including Clementine, which revolutionized our view of the Moon with a spacecraft that used Department of Defense technology then in development for the Star Wars missile defense program.

The Near Earth Asteroid Rendezvous mission, called NEAR, took only twenty-seven months to build and launch at a cost of $234 million (less than a tenth the cost of Curiosity). It intercepted the asteroid Eros, studied the surface from a distance of about 35 kilometers (22 miles), and then, in the spirit of risk taking, landed on it, although the spacecraft had never been designed for that. Two solar arrays served as legs and the spacecraft's body as a third contact point as it gently settled to the surface, 315 million kilometers (196 million miles) from Earth. The descent provided extraordinarily detailed images. Mark, then at Northwestern University, was on the team that interpreted its unexpected geology and published a series of papers in high-impact journals.

"It's an incredibly fascinating way to make a living, and sometimes I'm embarrassed to tell people I get paid to do this," he said.

But by the time those papers were published, in 2001, NASA had abandoned the Faster, Better, Cheaper philosophy. After seven years of success, a series of embarrassing failures overshadowed the early accomplishments, especially the loss of the Mars Climate Orbiter, a weather satellite for Mars launched in 1998, and its companion, the Mars Polar Lander, launched the next year. Both died because of software flaws that should have been detected. The painfully simple reason the orbiter crashed made it especially famous: a software engineer had mixed up English and metric units in the program that calculated small adjustments to the spacecraft's trajectory on the way to Mars, causing it to miss its orbit and burn up in the atmosphere.

Goldin had said it was OK to fail, but this wasn't what anyone had in mind. A lesson drawn by engineers, often repeated at NASA today, is that when it comes to faster, better, and cheaper, you can have any two, but not all three. The concept became a rule of the possible, called the Iron Triangle. Senior management had set unrealistic objectives and time and cost limits, requiring corner cutting that contributed to losing the spacecraft. In part, there simply weren't enough trained engineers and scientists working on the teams. People consistently put in sixty- to eighty-hour weeks without peers to support them or check their work.

But the true causes were deeper, according to the investigation report on the Mars Climate Orbiter and other analysis at that time. NASA really had not abandoned the top-down culture that made workers feel powerless. For example, Goldin's managers had set arbitrary time and cost limits from above. Mission teams themselves should have been allowed to set the parameters of the possible based on reality.

Software mistakes happen, but when people communicate they can find critical mistakes and fix them. It wasn't only that scientists and engineers had crazy workloads—that happens on successful projects, too—but they also were walled off in groups that didn't talk to one another, and management squelched their ability to speak up about problems. The investigation report said, "Project management should establish a policy and communicate it to all team members that they are empowered to forcefully and vigorously elevate concerns as high . . . as necessary to get attention."

NASA took a different lesson, accepting that the Faster, Better, Cheaper initiative was a failure, and became more risk averse and management directed. But outside analysts have since reconsidered the outcome of the concept. Even with its failures, it was a success. The sixteen missions attempted under the initiative, including nine successes and seven failures, were collectively less expensive than a single NASA flagship mission. For a bargain price, the nine successes delivered loads of data and answered many scientific questions, and did so much faster than NASA had typically done.

Some of the results from Faster, Better, Cheaper are still coming in.

The first Mars rover was the cute little Sojourner, which landed in 1997, after just three years of development. Intended to work for a week, it lasted three months and may still function, but it can't communicate with Earth through its base station. John Callas came up through Mars missions at JPL during the Faster, Better, Cheaper years, arriving in 1989 soon after getting his PhD in physics from Brown. He worked on spectacular failures and big successes. In 2000 he pulled together a science team for the Mars Exploration Rover mission. The Spirit and Opportunity rovers were launched within three years, evolving the technology pioneered on Sojourner, but were larger, more robust, and didn't need a base station on the planet.

For the bargain price of $400 million apiece, Spirit and Opportunity weren't designed with redundant systems—one soldered circuit goes bad and the rover fails—and that's why NASA sent two. But they exceeded all expectations. Designed to operate for three months in 2004, Spirit lasted until it got stuck in 2009. Opportunity is still exploring twelve years later, with periodic software upgrades. It has covered more than 40 kilometers (25 miles), more ground than any other off-Earth vehicle.

Curiosity evolved from Opportunity. Callas's team is a tenth the size of the Curiosity team, and the Opportunity rover is smaller, less capable, and has a computer on board about 1 percent as powerful as a smart phone. The day we visited, the Opportunity team was debating which rock to look at next. Like the Curiosity team, Callas's engineers send a batch of instructions daily. On Fridays, they send three days' worth, one day for driving and two for sensing, to cover the weekends. Callas gets text messages from the rover any time of day to let him know how it is doing.

The mission is open-ended. "We don't know what we're going to see, where we're going, and what we can find out about," Callas said.

The Mars Exploration Rover mission originally set out to "follow the water." Opportunity is a geologist. The follow-up mission, the Mars Science Laboratory, the Curiosity rover, does geochemistry, looking at organic compounds in the Martian environment. All along, the studies have focused on the hope for alien life.

NASA began looking for life on Mars in the 1970s. The chances of finding anything alive seem quite low at this point, unless ancient bacteria have somehow survived below the surface. The story of Mars appears to be one of increasing hostility to life. At one time, the planet had a thick atmosphere, large bodies of water, and rain. Opportunity has found clays that were laid down by water with an acidity suitable for life. But all that's gone, and the rovers have also found peroxide-like chemicals toxic enough to sterilize the surface of any life similar to what we know on Earth.

Now NASA studies the history of habitability on Mars, not the presence of life. Mars was probably livable in the past, but did anything live there? And what happened to those livable conditions?

The best current theory blames Mars's weak magnetic field. Earth's molten core, powered by radioactive elements, churns like a dynamo, inducing a magnetic field that surrounds our planet, partly shielding the atmosphere from charged particles from the Sun. You can imagine it like the defensive shields surrounding the Starship *Enterprise*. At night during an Alaskan winter you can sometimes see the solar wind hitting the atmosphere through a gap in the Earth's shield. The magnetic field channels streams of solar particles downward at the North and South Poles, where the Sun's ions impact the upper atmosphere and illuminate the swirling aurora. Atoms of oxygen glow green or red, depending on altitude, and nitrogen blue or purple.

In the distant past, Mars probably had a hot core that produced a magnetic field, too, but when the core cooled the planet dropped its shields and the atmosphere began taking direct hits from the Sun's continuous stream of electrons. The atmosphere ionized and was mostly stripped away. With an atmosphere remaining that is less than 1 percent as thick as Earth's, solar and cosmic radiation beat down directly on the surface, radiation capable of killing any life we know of. Extreme temperatures froze or vaporized the water. Water ice remains at the poles, but human beings couldn't live on Mars outside of pressurized, radiation-shielded shelters.

The rovers explore Mars to prepare for a possible human visit. But as decades pass and rovers become more capable, the need for humans to do science on Mars becomes less compelling. Today, any trained person is a lot smarter and more agile than a rover, but, as Callas said, "While they're trying to figure out how to keep humans healthy in space, these rovers are evolving."

Scientists see other places in the solar system as more likely to host life, but Mars research consumes by far the largest share of NASA's planetary science budget. A complex and expensive new mission is being planned to go to Mars, gather soil samples, and eventually bring them back to Earth for examination. Such a handful of dirt would cost billions but a lot less than sending an astronaut to look at the dirt there.

By the time we're ready for a space colony, Mars will be thoroughly understood. But we already know it's a difficult place to live. After half a century of exploration, what's remarkable is what we haven't found there. It seems to be neither a source of riches nor a refuge from environmental hazards on Earth. Other than its proximity to the Earth, there isn't a compelling reason for human beings to go to Mars.

When Amanda spoke in Tucson in 2014 at a meeting of the American Astronomical Society's Division for Planetary Sciences, a Chinese scientist ended her session with a long series of extremely detailed questions. They were the kind of questions that seemed to go beyond academic interest to a desire to learn how to replicate NASA's capabilities. The less-developed nations coming up behind the United States have the advantage that they can copy successes from the past. That's quite different from going somewhere new or trying something that has never been attempted before, which is what NASA is still doing in planetary science.

The pattern is familiar: science and technology leaders in the

West innovate and Asian countries with a different culture or lower costs of production reproduce those products cheaply and in numbers. Spacecraft components that were cutting-edge inventions for NASA in the last century became off-the-shelf equipment for companies launching satellites for communications, GPS navigation, and such. Private companies and smaller nations can reach space now. But the frontier of knowledge still belongs to NASA, the European Space Agency, and some lesser players, because discovery happens only once. It cannot be reverse engineered and replicated.

After being questioned so closely at the Tucson conference, Amanda checked in with a Chinese colleague to find out what was going on. His take was that the questions weren't part of any shadowy Chinese government directive, but the product of a scientist gathering details to make himself more valuable to the Chinese program.

Competition in space comes even more unexpectedly from India, but in 2014 the Indian Space Research Organisation put an orbiter around Mars using all-Indian technology. Prime Minister Narendra Modi bragged that the mission had not only cost less than an American space mission, it had cost less than an American movie about a space mission. At $74 million, the Mangalyaan orbiter actually did cost less than a lot of movies about space, including the 2000 Hollywood bomb *Mission to Mars*. India also plans to land on the Moon.

But these accomplishments are still remarkable mostly because they copy American space capabilities, not because they come close to exceeding them. India plans to put a rover on the Moon, like China and Russia before it. Google is also funding a competition, with a $30 million prize for the first private team that puts a rover on the Moon and sends data back, and various teams are competing from universities all over the world.

What will these missions learn? The Indian Mars mission's objective was to demonstrate that India could do it. The science objectives were secondary, adding peripherally to what NASA is already doing with more capable spacecraft. The science payoff from the Chinese lunar program is so far underwhelming. The program is technically strong but scientifically weak, with data analysis lagging and the expectation of new discoveries still vague.

We don't really know China's ultimate purpose in going to space.

The Chinese government is difficult for Americans to understand. It isn't a solid monolith that responds smoothly to dictatorial directives. Instead, many large bureaucratic organizations compete and cooperate in different spheres of control. Just because the public has no voice, that doesn't mean China doesn't have politics. It has office politics.

Chinese space scientists began proposing a Moon program in the 1990s, but their plans were repeatedly rejected by higher-level officials as lacking science objectives and being insufficiently thought out, a process documented by Patrick Besha, of George Washington University. The bureaucracy finally reached a consensus on a three-stage, multiyear lunar exploration plan in 2004. In 2007, the Chinese reached the Moon with the successful flight of the Chang'e 1 lunar orbiter.

The focus and vitality of the effort reflected the common cause felt by those working on it, the strong support of upper-level officials, and the program's youth—like Kennedy's Apollo program. From the perspective of the West, the orbiter seemed to spring from nowhere, but that was mostly because the Chinese had hidden their activities until they were certain of success.

Americans in general didn't know what to make of the Chinese space program. A cultural gap made it easy to ridicule. In 2013, when China's Jade Rabbit lunar rover ran into trouble hibernating before the fourteen-day lunar night, the government's Xinhua News Agency announced the problem with an odd first-person statement attributed to the rover itself. The Jade Rabbit said it might not survive the night and asked Chinese citizens to console its bereaved Chang'e 3 base station.

"The Sun has fallen, and the temperature is dropping so quickly," the rover was quoted as saying. "To tell you all a secret, I don't feel that sad. I was just in my own adventure story—and like every hero, I encountered a small problem. Goodnight, Earth. Goodnight, humanity."

Chinese followers of the Jade Rabbit sent back encouraging and heartfelt messages to the dying rover over the Chinese version of Twitter. In the United States, actor Patrick Stewart dressed in a foil rover suit and did a dramatic reading of the Jade Rabbit's last words on *The Daily Show*. The Chinese government, as always, provided scant technical details.

But it turned out that the Jade Rabbit did survive the night and kept operating long after its design life. In 2014, China demonstrated the ability to bring a lunar orbiter back to Earth as part of a multimission sample return program. Chinese astronauts had already flown twice in the Shenzhou orbiter atop Long March rockets. At this point, it seems likely that China will land astronauts on the Moon by its 2025 goal.

FUTURE

Late-night comedians in the United States were still making fun of the Chinese space program when it suddenly announced that a manned mission had lifted off and was on its way to the Moon, sending the first human beings beyond low Earth orbit since the Apollo program. The Chinese program had been developed in secret, with new capabilities coming at a rapid pace, incremental improvements that grew toward the ability to put human beings back on the Moon. The jokes were about how the Chinese were mastering 1960s technology, moving into a twentieth-century space age. American media adopted the attitude that China was like a younger brother, progressing through growth stages the United States had long passed.

When Chinese astronauts landed on the Moon, speaking a language that Americans could not understand, the laughter stopped. The Soviet Union terrified Americans in 1957 and electrified politicians to action by orbiting Sputnik, the first satellite, which started the space race. Their great-grandchildren felt similar emotions when China put people on the Moon. The accomplishment wasn't retro at all. The Chinese had done something the United States seemed to have lost the nerve to do: they focused on a stretch technological goal, invested in it, and made it happen.

From the outside, the Chinese program seemed to have advanced with shocking speed. In fact it had been developed over decades in secret, had been planned in exquisite detail, and was pursued with the single-minded resolve possible for a nondemocratic government.

The regime took a long time to decide to pursue a Moon mission, but once it adopted the plan, it didn't change its mind. When inevitable mistakes and accidents happened, officials didn't cringe and lose resolve due to the embarrassment and public criticism, because the mistakes were secret. The world never knew about the Chinese Moon missions that didn't make it.

The shock of being second best also made American leaders more willing to push the envelope of risk and spending. International competition had been the strongest motivator for human spaceflight in the past, and it would be again. Washington needed only weeks to choose an even harder goal than the Chinese, appropriate the first batch of money, and say the right words about risk and exploration. Americans would go to Mars. (The Moon had already been done, twice.)

During the crash program to build the American Mars spacecraft, China kept sending more astronauts to the Moon, building a habitat, gathering water from the ancient ice in permanently shadowed areas, and prospecting for minerals. But, similar to the pattern of the Apollo missions in the twentieth century, public excitement waned as the work became more routine. Chinese scientists reported interesting facts about the Moon, but nothing that could move the stock market. There wasn't an economic reason for Moon exploration or any practical way that astronauts could stay there without constant resupply from Earth. The habitat was like a space station on solid ground, and space stations had been done. Fun facts about the Moon were not enough to drive investment, and the Chinese began drawing down their presence.

Seeing those results, and looking at the long history of exploration of Mars by orbiters, landers, and rovers, NASA designed a mission to go, collect some rocks, and return. This would be a sortie mission to demonstrate American pre-eminence in space. Planners established nominal science goals, too, but there weren't many major scientific questions about Mars that rovers and landers hadn't already settled. The mission website asked if there were oceans in Mars's past. Although already established and re-established many times since 1965, detecting the presence of water still sounded like an interesting research goal to put in the publicity material.

The far greater cost of leaving people on Mars, resupplying them, and building a colony would have to wait for future decisions and investment. When it came down to spending the kind of money a permanent Martian outpost would require, political leaders needed a return on investment. The cost was just too great to put people there with nothing for them to do but survive. No one had found a way to make a profit on Mars, and Mars did not look like a refuge from anything bad that could happen on Earth. Even an apocalyptic Earth would be safer than Mars. If the Earth were bathed in radiation, poisoned in its atmosphere, and stripped of life-supporting resources, Mars would still be worse.

In the foreseeable future, there didn't seem to be any better reason to provide a home on Mars than on the Moon. Human beings could be placed there and supported by spacecraft ferrying their supplies from the Earth. That would be more like camping in the backyard than heading off to build a home on a new world.

So the Chinese conquered the Moon, set up a camp, and came back. The Americans planned to do the same on Mars. Would-be space colonists studied their work and thought about applying it to the far more challenging work of leaving the Earth for good.

PRESENT

NASA began sending probes to other planets at the dawn of the space age, partly in search of an Earthlike world. Carl Sagan lobbied NASA to send a spacecraft to Venus soon after the agency formed, and Mariner 2 made it in 1962, learning that the atmosphere was hotter than a baking oven. The average surface temperature on Venus is 465 degrees Celsius (870 degrees Fahrenheit), hot enough to melt lead. More than twenty spacecraft have followed to fill in the picture of the planet that is sometimes called Earth's twin.

Before the probes gathered their readings, Venus had looked like a good destination for people. It's about the same size and has the same gravity as Earth, it is Earth's closest planetary neighbor, and its thick atmosphere would protect the surface from radiation. But the atmosphere is much too thick, much too hot, and contains corrosive sulfuric acid, and the winds at high altitudes are ferocious. At the surface, the atmosphere is as heavy as the deep sea on Earth, and even gentle winds move objects the way water currents do in our oceans. The Soviet Union tried to land a series of probes on Venus. Those that survived the descent through its hellish atmosphere lasted only briefly on the surface. Human beings couldn't go there.

But missions like the European Space Agency's Venus Express found evidence that Venus probably was once a lot more like the Earth, with oceans of water and a more reasonable atmosphere. Venus's weak magnetic field allowed the solar wind to strip away most of the water in the first billion years of the planet's existence. Earth and Venus still have about the same amount of carbon dioxide, but on the Earth the oceans have gathered it up and converted it into limestone and other geological deposits. On Venus, without water to do that work, the carbon dioxide remains in the atmosphere. Venus receives more sunlight than the Earth, and its atmosphere of 97 percent carbon dioxide creates a heat-trapping blanket that fries the planet. Over the last century, human beings have been making the Earth more like Venus, releasing carbon dioxide from our geology back to our atmosphere and warming the planet (although our atmosphere is still only four-hundredths of a percent carbon dioxide).

Mercury, the least studied of the four rocky inner planets (the others are Earth, Mars, and Venus) is the weirdest and least appealing to visit. Mercury is tiny, lacks an atmosphere, and is dominated by its close proximity to the Sun. One side is blazing hot, the other side extremely cold. A visitor would be hard-pressed to come back, fighting the Sun's gravity on the return trip to Earth. NASA sent the Mariner 10 probe in 1973 and the MESSENGER orbiter in 2004, which ended its successful mission by crashing into the planet in 2015.

Decades of exploration have told us most of what we need to know about colonizing the inner solar system. The prospects are not good for people thriving on their own on one of these bodies, unless we could transform Mars or the Moon into a habitable environment. That idea, first advanced by Sagan in the 1970s, entered the scientific literature in the 1980s with the word "terraforming" in a paper by

Chris McKay, now a planetary scientist at NASA's Ames Research Center.

Like a lot of ideas about space, popular culture changed terraforming into an imaginary process far easier than it could really be. McKay batted down one of those notions, the plan of blasting Mars's ice caps with nuclear weapons to vaporize them and make an atmosphere. McKay said all the world's nuclear weapons contain energy comparable to less than five hours of sunlight hitting Mars, and most of that would be lost in huge explosions. It would take far more energy for a far longer time to make a difference.

But McKay's ideas do rely on making an atmosphere. He would mine Mars to produce superstrong greenhouse gases called chlorofluorocarbons, or CFCs, which would blanket the Martian atmosphere as greenhouses gases do on Earth, warming the surface to melt the ice caps. Carbon dioxide released by that melting would further increase the warming. Producing CFCs on this scale would require a massive industrial process powered by vast amounts of energy— more than 25,000 times more CFCs than annually were produced on Earth. And it would take about one hundred years before that process would warm Mars and create a CO_2 atmosphere. After that, growing planet-spanning forests and fields could produce breathable oxygen on Mars in about one hundred thousand years.

These are fun ideas to think about, but the time and cost are too colossal to take seriously. We doubt people would invest money in power plants, robotic mines and factories on another planet that would not create a habitable colony within anyone's lifespan. We don't know if anything would grow on Mars. And until the atmosphere could protect humans from radiation, the planet would be available only for short visits or stays underground. Robots would have to do all the work.

If the problem driving us to Mars were a deteriorating environment on Earth, it would probably make more sense to use the wealth, energy technology, and innovation to fix this planet. That task is technically and economically far easier.

We've ruled out the inner planets other than Earth as places for a permanent human colony. Half a century of exploration gave us that information, but it also did much more than that. It built the ability of NASA and other space agencies to innovate and reach for new kinds of knowledge. While manned spaceflight stalled, planetary science grew and thrived. As this is written, Earth has a couple of dozen active space missions in flight, with more launched every year.

These lessons are leading to new missions that are more ambitious, go farther, and promise to find a place where humanity can make a new home, beyond the inner solar system.

3

A HOME IN THE OUTER SOLAR SYSTEM

Amanda knows of a place for a space colony, orbiting Saturn. Other people like it, too. Instruments on the Cassini probe she works on detected signs of liquid methane in the lakes of Titan, Saturn's moon, and now there are hopes to send a boat or submarine to study those depths. Maybe methane-based fish are swimming there.

Amanda's search began at age seven, riding across the Mohave Desert in her father's VW van, when she saw the bottomless darkness and dizzying light of the unobstructed night sky. She couldn't forget the feeling of that immense and brilliant vision. She wanted to go there. In third grade, a young student teacher named Holly got her on the way with a lesson on the solar system. Amanda bonded with Holly and her love of the planets. She made a diorama of the solar system out of a Stride Rite shoebox. She still treasures the model, with its painted clay planets hanging from strings, including Pluto.

NASA landed Viking on Mars that year, the first spacecraft to safely set down on another planet. On the front page of the newspaper Amanda saw the reddish sky of Mars, the rocks and sand on the surface, and the horizon, another planet to explore. She became focused on space, going there or at least working at the Jet Propulsion

Lab, where the spaceships came from. It was near her house in Pasadena but mysterious. No one in her family had followed such a path.

As a teen, Amanda drove her mom's car to science talks at Caltech and Pasadena City College. In high school she already planned to get a PhD in a space-related field. She visited Caltech in January 1986 for a public event as Voyager 2 flew by Uranus (a few days before the *Challenger* disaster).

The Voyager missions had been launched in 1977 to fly by Jupiter and Saturn, including Saturn's moon Titan, a mission planned to last five years, but with the hope the probes could go farther. The four outer planets were aligned, an event that would not recur for 175 years. NASA extended the mission beyond Saturn to Uranus and Neptune. In 1989, they flew beyond the orbit of Pluto. The Voyagers are still flying. Voyager 1 is beyond the solar system, in interstellar space, and Voyager 2 will follow soon. They still send back data, although it takes more than a day for a signal traveling at the speed of light to make it to Earth and back.

The Voyagers won't approach another planetary system for at least forty thousand years (Voyager 1 is headed toward a star in the constellation Ursa Minor, the Little Dipper; Voyager 2 in the direction of Andromeda). Carl Sagan chaired an international committee to produce messages on board for aliens on planets around other stars. People of a certain age can hear his voice in their memories and see his turtleneck shirt, inspiring us about the billions and billions of kilometers the Voyagers would carry these messages. So far they've gone about 20 billion kilometers (12 billion miles).

Galileo followed up on Voyager to visit Jupiter. It's hard to build probes to the outer solar system—they take a long time to get there, and the loss of the *Challenger* added delay—so by the time Galileo was on its way, Amanda was done with high school and college and was looking for a topic for her doctoral dissertation at the University of Colorado. Justin Maki was a classmate, already working on Mars, and suggested she call Dr. Charles Barth, a Colorado professor who had started his space science career at JPL in 1959, had run the ultraviolet instruments on Mariner 6, 7, and 9 to Mars, and was currently involved with the ultraviolet instrument on Galileo. Barth said, "Come on over, we can use you."

Barth showed Amanda the first image that came back from Mariner 4, in 1965. Mariner 3 had failed and NASA's Richard Grumm, impatient to see if the Mariner 4 system had worked, decided to take the binary data coming back, print it out on ticker tape, staple the tape to a wall, and color it in, paint-by-numbers style, using pastels he'd picked up at a Pasadena art shop (the pastel set is now in a museum). The exercise worked, and the picture, later preserved by cutting out that wall section, is the first image ever taken from a vehicle off the Earth.

The science project Barth gave to Amanda was a bit more sophisticated. Galileo, on its way to Jupiter, had pointed its ultraviolet instrument at the Moon as it passed by, but no one had analyzed the data. UV had been used mostly to look at atmospheres, not surfaces, and the value of the data was unknown. Making sense of it required figuring out how to interpret spectral signatures from the Moon's surface. By comparing those lines with similar squiggly lines from a detector looking at samples in the lab, Amanda could identify materials the probe was seeing on the lunar surface. Since UV doesn't penetrate far into the surface, the data were particularly good for looking at how radiation from space had transformed the Moon's dusty soil.

Galileo was still on the way to Jupiter when the usefulness of this work became clear. During the planning phase of the mission, long before Amanda's career began, designers assumed the UV detector wouldn't work in the high-radiation environment near Europa, one of Jupiter's moons. Amanda and her colleagues believed differently. With her new PhD in hand, she was dispatched to a Brown University meeting of scientists studying Galileo's icy satellites to pitch the use of the UV sensor to look at Europa. Her proposal put her up against teams for each of Jupiter's other instruments, who all wanted their own investigations to use the probe's scarce and precious data-gathering time. Galileo had only hours to observe Europa during the flyby, and its high-gain antenna had failed, squeezing the capacity of data that could be recorded and transmitted.

The competition was intense. Amanda remembers being totally unprepared for the aggressive attack she endured. She felt she had been destroyed by her challengers. But the chair of the meeting heard. Time for the UV spectrometer was approved and it did gather valuable data at the first Europa flyby, in December 1997. Europa turned out to be a strange place, a big icy ball with a surface that looked, close up, like a frozen lake.

Cassini also followed up on Voyager's discoveries, bound for Saturn rather than Jupiter. It was conceived in the early 1980s, launched in 1997, and arrived at Saturn in 2004. By that time, Amanda was a mature scientist and co-investigator on the ultraviolet instrument, working in the windowless Space Flight Operations Facility at the Jet Propulsion Lab in Pasadena—the SFOF at JPL, to use the ubiquitous acronyms of space exploration.

When Amanda had visited JPL during the Galileo mission, Justin Maki had shown her around a maze of cubicles where scientists and engineers were planning Mars Pathfinder's observational sequences. When she came back to work on Cassini, a decade later, she found the same cubicle maze. But now it had been rearranged with different pathways. Charts and artwork about Saturn and Cassini covered the walls and cubicles and hung from the ceiling. It was Cassiniland. Nerds become cool when they work here, and their offices were decorated with stuff from *Star Wars* and other pop culture fantasies.

A good planetary scientist is a team player, an optimist, and gifted at delaying gratification. Missions to the outer solar system take decades to plan, design, and, as funding and priorities change, to replan and redesign, before they make it to the launchpad—if they ever do get that far. Amanda and her colleagues have dedicated years of effort, imagination, and hope to spacecraft that never got built, or blew up on the launchpad, got lost in space and stopped communicating, or crashed into another planet.

With so much of each scientist's working life on the line, moments of peril for these spacecraft are shared agony among many colleagues. The hours when Cassini arrived at Saturn in 2004 were like that. Everyone gathered near the mission control room on the ground floor of the SFOF with an overflow crowd in a large auditorium at nearby Pasadena City College. Glowing screens cut the darkness of the control room with colored numbers and images showing the status of each of the spacecraft's systems. To slow down its flight and enter Saturn's orbit, Cassini would have to round the planet, going out of radio contact, fire its large rocket, and pass through the plane of Saturn's rings—which could contain particles that would destroy it.

The radio silence would last nearly an hour. There would be no way of knowing if the maneuver was successful other than waiting for the signal from Cassini to resume. If none did, that meant that decades of work and the hopes of hundreds of scientists would disappear into empty space—half of a two-mission career for many of them. When the hour passed, Cassini signaled that it had survived and entered orbit, and the relief and celebration were overwhelming.

Compared to a moment like that, moments of discovery sneak up on people at JPL. Amanda has been deeply involved in looking at Saturn's moons and has authored a series of articles in the world's top journals on her discoveries. When something interesting was seen, news filtered through the hallways and across the e-mail of researchers working in the building and at institutions around the globe—the team is virtual, always connected by conference calls and online. No one would say anything publicly, to avoid scooping the news, but privately everyone would feel the vibration of an exciting new finding.

Before Cassini's arrival, the orange atmosphere of one of Saturn's moons, Titan, hid its surface from view. Only radar and some infrared wavelengths can penetrate the haze. Scientists speculated that they might find oceans of ethane or methane—hydrocarbons like natural gas on Earth, but so cold they would be liquid. On arrival, however, Cassini wasn't picking up anything like that. It dropped the Huygens probe to the surface (both Cassini and Huygens are named for seventeenth-century astronomers who discovered moons of Saturn). Huygens had been designed to float and measure the size of waves, but it fell on damp, soft ground in an area scattered with pebble-like chunks of water ice.

When Cassini flew over Titan's polar regions, however, its radar picked up something that looked smooth, like a lake. Around the edges of this smooth area were branching shapes that looked exactly like the channels, bays, and coves of a shoreline on Earth. Another instrument on board could measure reflected sunlight. At the right moment, when the sunlight would be bouncing off that possible lake, Cassini looked in the direction where it would be expected to be glinting off liquid. To Amanda's delight, it looked exactly like afternoon light reflecting off lake waters on Earth.

Titan has the only surface liquids in the solar system other than the Earth. Its enormous lakes hold many times more hydrocarbons than have ever been discovered on our planet. Cassini's gravitational measurements suggest that a slushy ocean of water lies within Titan, but the clouds, rain, rivers, and lakes on the surface are liquid ethane and methane, like the contents of liquefied natural gas tankers. Titan has weather, beaches, and tides, but it is colder than a deep freeze. It's a familiar environment and a weird one at the same time.

And there's much more left to learn. When Amanda or one of her colleagues develops an idea—a hypothesis about Titan—he or she can propose to the team that Cassini take a look. But patience and optimism are required. If the measurement hasn't already been planned into the trajectory of the spacecraft, scientists and engineers must gather to consider the cost in fuel and the risks of the maneuver against the value of the information to be gained.

It takes an hour and a half for a command to get from the Earth to Cassini. Mistakes can't be fixed quickly, or maybe at all. Once a measurement has been approved, every command is modeled on Earth to check for safety to the spacecraft. At least several months will have passed, and sometimes several years, from a scientist's idea to receiving data.

The process is slow, but it works over time. A surprisingly detailed view of Jupiter, Saturn, and their moons has emerged. It's a menagerie of strange worlds, by far the most interesting places to study in the solar system. Europa's crust of ice covers liquid water. Scientists found that hidden ocean by measuring an electric current from it, which was induced by Jupiter's magnetic field. Saturn's Enceladus sprays giant geysers of water vapor and ice into space from its south pole. Along with Jupiter's Io and Ganymede, the warmth inside these moons comes from the constant flex and crunch caused by tidal forces from the intense gravitational fields of Jupiter and Saturn.

What we already know about the solar system suggests we look outward, to these moons, for a place to colonize.

The planets were born of a disk of dust and gases orbiting the Sun. While still floating freely, the heavier elements condensed in the warmer regions toward the center of the system. When the planets coalesced, like lumps in this soup, those drawing material from nearer the Sun formed of rock and metal. Planets in the outer solar system gathered up the lighter elements and are mostly made of ice and gas.

Water in liquid and solid forms is abundant far from the Sun. The moons of Saturn and Jupiter have rock inside them, but water makes up much more of their volume than on the rocky inner planets. For example, Titan is larger than the planet Mercury and has a radius 50 percent larger than Earth's Moon, but its density is less than either, and so its gravity is weaker. Water is less dense than rock and metal.

Jupiter has some sixty-seven moons, four of which were discovered by Galileo (the man, not the probe), and are large enough to consider for a colony: Ganymede, Callisto, Io, and Europa. Scientists believe the smallest of these four, Europa, is one of the most likely places in the solar system to find life, because of its liquid subsurface ocean. It must be dark down there, so the Sun wouldn't power life, as on Earth. But a few unique forms of life on Earth thrive in eternal darkness beneath deep oceans and underground, drawing energy from sources other than the Sun, so that could happen on Europa, too.

How thick is the ice on Europa and could we get through it to see if anything is swimming around? From the Galileo data, scientists believe the frozen layer is 10 to 100 kilometers thick (6 to 62 miles), but there also seem to be places where bergs are at the surface, suggesting thin ice, and maybe even areas where plumes of warm water break through. A group from the Southwest Research Institute, in San Antonio, used the Hubble Space Telescope to pick up ultraviolet data from Europa suggestive of a spewing plume of water, which, if real, would make it much easier to find out what's going on underneath. But that result has not been duplicated and is controversial.

In 2014, NASA called for proposals for a mission to learn more about Europa, including investigating the plume. Amanda put together a team with the University of Colorado's Laboratory for Atmospheric and Space Physics, where she began her career, to build an ultraviolet instrument for a Europa probe. With just ninety days to create the scientific justification, design the instrument, and develop cost details, the work became an intense rush of conference calls with scientists and engineers from many institutions. They were competing with many other teams who wanted to put their own detectors and cameras on the possible probe.

Competition is constant in planetary science. Competition for missions, to put instruments on missions, and for time to operate instruments when they reach their destination. It seems to have worked, as planetary science has developed and expanded while human spaceflight, run from inside NASA, stalled years ago.

Some ideas for managing science have survived from NASA's Faster, Better, Cheaper initiative in the 1990s. NASA lets universities and outside groups compete to create and operate missions that are smaller, less expensive, and get launched more quickly than its flagship missions: the low-cost Discovery Program and midrange New Frontiers Program. Institutions compete to design the missions under preset budget limits, so the goals stay realistic. JPL is part of many of these competitions, scrubbing each proposal through an intense analysis by a cool and prestigious internal group called "Team X," which scrutinizes all aspects of the mission, from the trajectory to navigation to science goals to data downlink.

NASA itself still develops and operates the big, multi-billion-dollar missions like Viking, Voyager, the current Mars program, and outer solar system missions such as Galileo and Cassini, which carry instruments developed through outside proposals. These are called flagship missions, and they only seem to get off the ground every decade or two. It's tough to get to Jupiter or Saturn without building a flagship mission. The spacecraft need a higher level of sophistication to make the seven-year journey, with internal power supplied by a lump of plutonium, as the Sun is too dim that far out to use solar cells.

Several ideas for looking at Titan could peel back its orange clouds to find out how its complicated weather and geography work, how easily humans could live there, and if anything lives there now. Everything about Titan is both strange and familiar: it is a world made of hydrocarbons that mirrors our world of water. Rain, seasons, waves, dunes, bedrock: all are present on Titan, but all are made of different chemistry than their analogs on our own planet. Simple curiosity demands we find out what's going on there.

The leading idea has been a grand and complicated spacecraft called the Titan Saturn System Mission, the TSSM, which would include an orbiter carrying eight instruments, a balloon to float in Titan's atmosphere carrying eight more instruments, and a boat that would float in a northern lake with five instruments. The balloon and the spacecraft itself would be nuclear powered, with heat and electricity from plutonium. But the mission would get to Titan with a new technology, an electric motor called a Hall thruster powered by solar energy.

There also has been a proposed boat-only mission called the Titan Mare Explorer that could be done much less expensively, as a Discovery project. And another idea calls for a series of landers that would observe Titan's surface in several of its enigmatically different climatic and geologic zones.

None of these plans is anywhere close to launch, and they certainly won't all go. Big, complex missions tend to get bigger and more complex until they get scaled back and redesigned or canceled. The process is political, bureaucratic, competitive, and uncertain. But somewhere, engineers are developing the ideas and taking them closer to reality.

"We have always known that it is intensely competitive," said Julian Nott, who designed a balloon for Titan. "Will your ideas get selected for a Titan mission? Answer: probably one in ten. Not much chance. But maybe you'll propose ideas and somebody else will take them up."

For now, Cassini is still sending back valuable information and will keep operating until 2017. There had been hope of launching a Cassini follow-on mission by 2023, but that appears unrealistic now. If such a spacecraft launches in 2030, it won't arrive until 2037 (unless a faster rocket shortens the trip). By that time, the scientists who began their careers as graduate students on Voyager will have retired. Amanda should still be working, however, and she's looking forward to seeing what the data say, twenty-one years from now, if such a mission is successful.

At NASA's current pace, degradation of the Earth is happening much faster than discovery of a possible new home. But the pace isn't dictated by science. As in NASA's record of manned missions, the agency moved faster in planetary science in the 1960s and '70s, spending more and launching missions more frequently. To make faster progress, we need more money and more ambition. Exploration plans could be implemented in years rather than decades. Missions could launch before the previous mission arrives. A more motivated Earth could speed up the process now and get answers from Titan much quicker.

Alone in the solar system, Titan's landscape is buried in fuel we could harvest and burn with technology hardly more advanced than the gas furnace found in a typical American house. The natural gas on Earth is mostly methane, like Titan's lakes and seas. The dunes around those shores on Titan are also hydrocarbons, mostly of heavier and more complex organics called polycyclic aromatic hydrocarbons, or PAHs. With Titan's atmospheric hydrocarbon factory and cold temperatures, this all makes sense.

So why doesn't Titan explode when someone lights a match? No oxygen. We burn fossil fuels on Earth by adding heat or a spark to a combination of carbon-rich fuel and oxygen. Energy comes out in a flame or explosion as the energy originally deposited by the Sun is released, along with carbon dioxide and water. Titan's atmosphere is mostly nitrogen, like Earth's, but without oxygen.

But underneath Titan's hydrocarbon surface layer, perhaps just below, or maybe 100 kilometers (62 miles) down, water ice or slush makes up much of its mass. Water contains plenty of oxygen. A simple process of running water through an electrical field, called electrolysis, frees the oxygen. The International Space Station uses electrolysis to produce oxygen for breathing. The colonists could also breathe it, and could use it to burn methane, which would provide plenty of energy to keep the process going.

Explorers could go to Titan with their own energy source—say, a small nuclear reactor—and make it their first priority to mine the underground water supply and separate the oxygen with electrolysis. The energy recovered from burning the methane and oxygen would be more than adequate to power further ice excavation, electrolysis, and the heating of a habitat, along with everything else the colony needed.

With Titan power plants running on hydrocarbon fuels, colonists could build large, lighted greenhouses to grow food and process the carbon dioxide exhaust from fuel combustion back to oxygen. With the materials at hand they could build almost everything out of plastic. For metals and other heavier elements needed for nutrients or to make electronics, the colony could dispatch spacecraft to mine asteroids. With limitless energy and access to resources, colonists could eventually build lakeside homes, go boating, and fly personal aircraft.

Many scientists have imagined what it would be like to live on Titan, because it seems so easy. Ralph Lorenz of the Johns Hopkins Applied Physics Laboratory has written a couple of books about Titan. He has proposed various exploratory missions, including a boat, like a buoy, and a set of weather stations. And when we spoke he was thinking about a submarine. "There is no vehicle on Earth that you cannot make sense of using somewhere on Titan," Ralph said.

Lorenz points out that Titan is a place where human beings could survive without spacesuits, walking around in warm clothing and oxygen masks, and could live in nonpressurized buildings. It's not that hard to imagine standing on Titan's strange, orange landscape, on damp, soft ground such as the Huygens probe encountered, with pebbles of hard ice here and there. The temperature is around -180 degrees Celsius (-290 degrees Fahrenheit), but clothing with thick insulation or heating elements would keep you comfortable. A rip wouldn't kill you as long as you didn't freeze. You wouldn't need a bulky pressurized suit, like the ones worn by astronauts on the Moon or in the vacuum of space.

A habitat on Titan could be built with a design similar to those used in polar environments on Earth, with plenty of airtight insulation and legs to keep the warm interior from melting the frozen water and hydrocarbons upon which buildings stand. A simple double door would keep the oxygen inside. If the habitat sprung a leak, someone would have to fix it, but there would be no immediate danger. A piece of tape might take care of the problem until a proper repair. The ubiquitous hydrocarbons contain many carcinogens, so it would be important to clean up when coming indoors and taking off outdoor gear.

Titan and Antarctica have some similarities. To survive in either place requires active technology, most importantly for heat. Both environments call for all supplies to be brought in. To stay permanently without support, you would need a source of energy and indoor food production. Antarctica probably has plenty of fossil fuels, although any mining or drilling would have to go through thick ice to get to them. On Titan, fuels are common on the surface, but oxygen would have to be mined or drilled from underneath. To go outdoors at either place, you must get properly dressed. While the outside temperature is much colder on Titan, the weather is calmer.

The ability to breathe the air freely in Antarctica is a major difference from Titan. The atmosphere on Earth is close to 80 percent nitrogen and 20 percent oxygen. Titan's is 95 percent nitrogen and 5 percent methane. We can't survive without oxygen to breathe, but Titan's air wouldn't be instantly poisonous, either. There's enough cyanide to give a person a bad headache and the nitrogen would cause narcosis, as it does with deep-sea divers, a reversible condition similar to drunkenness. If you lost your breathing apparatus, you would pass out within a minute, but you could be revived with oxygen provided in time.

The pressure of Titan's atmosphere is 50 percent greater than Earth's. The atmosphere is more than adequate to provide protection from radiation or micrometeoroids. Because of the cold, the air also is four times denser than on Earth. That has two curious side effects. One is the weather's stability and slow change. The other is that people can fly with very little difficulty in Titan's low-gravity environment.

Titan has only 14 percent of the gravity of the Earth, even less than our moon, where gravity is 17 percent of the Earth's. (Titan is much larger than the Moon, but the Moon contains more rock, which has greater gravity-generating mass than the water that makes up most of Titan.) In the Moon's low gravity, the Apollo astronauts got around with a hopping movement, seeming to jump in slow motion like party balloons bouncing along the floor. On Titan, with less gravity to hold people down, they would also be supported aloft by a thick atmosphere: with winged suits, they could effortlessly glide great distances.

Adding just a little propulsion, humans could fly on Titan. You could do it by flapping wings attached to your arms, or with a system like a pedal boat to turn a propeller. An electric-powered prop would be more practical and comfortable, because vigorous exercise in heavy insulated clothing can be uncomfortably soggy. If the wings fell off, the flier would float to the ground at about 15 miles per hour. Terminal velocity—the maximum rate at which objects fall in the atmosphere—is a tenth as fast on Titan as on the Earth.

An even more dramatic difference between life on Antarctica and Titan is in the ability to go home again. The human body would probably adapt to Titan in ways that would make return to Earth difficult.

Gravity conditions our bodies. Runners develop stronger bones due to the forces of their feet colliding with the ground. Patients confined to hospital beds for long periods lose muscle tone and can become too weak to stand. NASA has figured out how to exercise astronauts on the International Space Station to help them retain muscle mass and bone density during six-month weightless flights, but that routine requires two hours of training a day in special machines. Most Titan colonists probably wouldn't keep up their workout routines any better than the typical Earth resident does with an unused gym membership. They would probably lose strength over time. Likely, they would become too weak to live on Earth.

Colonists also would have to rely on artificial light. Everyone who has lived at a northern latitude knows that natural light and darkness regulate life, moods, and the ability to function both outdoors and in. At the Earth's poles, the Sun is up all summer and the sky is dark all winter. No one lives at the poles other than researchers, but residents of northern regions far south of the pole still must adjust to the variations in light physically and with technology. Indigenous people hunkered down in winter and relied on foods such as marine mammal oil and organ meats for the vitamin D that residents of temperate climes get from the Sun. In the summer, people in the north become busy and energetic, storing up food over long sunny days.

Modern residents of polar climates rely on indoor light sources to maintain their daily cycle of sleep and waking. They eat processed foods that provide vitamin D (although often not enough). Without a regulated daily cycle and adequate bright light and vitamin D, many people become depressed, getting winter blues that begin when natural light decreases in the fall.

On Titan, indoor lighting and proper diet would be a year-round need. The natural cycles of light and darkness would be completely unfamiliar. As a moon of Saturn, Titan is tidally locked to the planet, with one side always facing it. However, the orange atmosphere probably blocks any clear view of the stars or planets. (In any event, Titan is on the same plane as Saturn's rings, so they would not be visible.) On Titan's Saturn-facing side, where a colony surely would be built, the constant glow of Saturn's reflected light probably keeps the sky dimly illuminated all day except when Titan falls in Saturn's shadow. The day lasts sixteen Earth days, so for a couple of weeks the Sun would add its weak light, and for another two weeks the landscape would dim. The Titan year is twenty-nine Earth years, so the four seasons are roughly seven years each. Cassini has studied Titan for nearly half of a Titan year, starting with summer in the south and now moving into summer in the north, but we are only beginning to understand seasonal effects on weather.

There's plenty we don't currently know about Titan. But we do know that, if we could get there, we could live there.

FUTURE

Around the world, the climate crisis unfolded with endless speculation about colonizing another planet somewhere in space, creating wildly unrealistic expectations about how many people might be able to leave the Earth. Twenty-four-hour news coverage of seacoast disasters and desert migration episodes often included an expert or two speculating on the livability of the moons of Saturn and Jupiter.

A ministerial-level meeting finally convened in Geneva to choose a destination. A hurricane had just eliminated the last of the East Coast's barrier islands in New Jersey and North Carolina, sending waves crashing into inland towns. In New York, waves had overwashed the partially completed Brooklyn dike, finishing off neighborhoods near Coney Island and Brighton Beach and flooding the

B, D, and Q subway lines. But such disasters were routine. News coverage focused on the meeting in Geneva, which seemed to be a source of hope.

Everyone in the ornate hall knew the choices well, but no seats remained for the briefing. Closely held new data had arrived from the outer solar system. This meeting would reveal whatever surprise it held. The International Commission on Planetary Prioritization (ICPP) had worked in secret.

The professor chairing the ICPP's Technical Committee couldn't suppress a smile of self-satisfaction as he moved to the lectern. A famous genius, Nobel winner, and media darling known for his wicked humor, he had the status and the ego to stand before an array of ministerial officials from global powers with the patronizing air that he adopted with undergraduates in his enormous lecture classes.

The professor decided to extend his moment in the spotlight by reviewing background information already familiar to the gathered ministers, secretaries, and presidential science advisers. He reminded them that the committee's mandate had been to find a new world meeting a fourfold test: (1) the colony would be livable for families and safe for reproduction; (2) the colony would not be at risk of catastrophic events that could wipe it out; (3) the colony would be economically rational as a long-term investment; and (4) the colony would be capable of becoming self-supporting should it lose contact with Earth.

"I think we are all aware that Mars and the Moon aren't suitable for building a self-sustaining colony," he said. "The advantages are that they are close and well understood. We can get there. But without an atmosphere, we would always have to live behind shielding or underground in a pressurized habitat. As we've learned, punctures in this kind of enclosure can lead to catastrophe, and excursions outside in spacesuits will always be complex and risky.

"Even if we defeat that problem, the evidence from our social scientists tells us that human beings don't want to live permanently underground. We can do that on Earth. We've seen underground subdivisions built for some very wealthy individuals, for shelter from climate and radiation events, and those facilities are usually unoccupied.

"Water on the Moon is available, but finite and complicated to harvest. The Chinese have built a facility to excavate ice from the lunar poles and produce hydrogen and oxygen through electrolysis powered by solar arrays, but the use of that facility has been limited to supporting the Moon missions themselves. Water on Mars would be similarly difficult to harvest.

"We've got some corporate interest in energy generation on the Moon with a larger facility like the Chinese base, using solar panels to produce high-density fuel that could be flown back to Earth. We're not sure if the economics will ever favor that idea, or if it might make more sense to do the whole process in orbit and beam power to the Earth using lasers or microwaves. In any event, energy remains too cheap on Earth to make such an investment viable for the foreseeable future.

"The surface of the Moon also has deposits of a helium isotope called helium-3, which is appealing for use in nuclear fusion reactors. Unfortunately, we're still a few years short of building a commercial nuclear fusion reactor that produces more energy than it consumes, and the reactors we know how to build are big and finicky. They can't go mobile anytime soon. If we ever do get fusion to work, we could harvest helium-3 and bring it back from the Moon quite profitably.

"But that would only meet one of our criteria. If resource extraction does become a working financial model for the Moon, it probably won't lead to colonization. It would be easier to move workers back and forth rather than relocate their families three days away to the Moon. And we don't see any prospect that a Moon base would ever be self-sufficient.

"Mars has plenty of frozen water, but the lack of evidence of life is disturbing. We're pretty sure there are good reasons why it is sterile. Also, we don't see an energy source. Bottom line, Mars is no more attractive than the Moon, and it is a lot harder to get there, as we're finding with our manned mission program from NASA.

"The next relatively close possible destination is Venus. We sent our last probe there decades ago. It has plenty of atmosphere. In fact, the atmospheric pressure is equivalent to 3,000 feet deep in Earth's ocean. It's also poisoned with acid and hot enough to melt lead.

"Those of you involved in the ongoing global warming debate

may be interested in why it is so hot on Venus: the atmosphere is thick with carbon dioxide. Venus cooks under a runaway greenhouse effect, the endgame of the same phenomenon we are experiencing here on the Earth."

The professor now read from a card that had been given to him by the public affairs group, in a mock-serious voice.

"However, I have been instructed to emphasize that we don't know for certain if the carbon dioxide we are emitting from fossil fuel combustion will be sufficient to make Earth uninhabitable like Venus. More study is needed."

He returned to his remarks.

"The other choices for a colony are farther away, in the outer solar system. The Sun is dimmer and the planets are gaseous. We can't build a space station on Jupiter or Saturn, because those planets have no solid ground upon which to build. They're more like suns that weren't large enough to flip on: all atmosphere.

"We've given some study to the idea of a colony in the atmosphere of Saturn or Jupiter. A habitat with the right buoyancy could bob at a predetermined level in the atmosphere of one of these gaseous planets, like a boat, providing appropriate gravity and a stable platform for living.

"This idea meets most of our planet-selection criteria: protection from radiation and micrometeoroids, exterior gas pressure, so no explosive decompression, an atmosphere that could provide some needed materials. We can harvest helium-3 there, as on the Moon, so that's a future benefit. But we would have no current source of power or heavier materials."

The U.S. secretary of state, sitting in the front row, cleared his throat noisily.

"Move on, Professor. No one wants to live in a boat on Jupiter. Let's get to the meat of the report: the moons of Saturn and Jupiter."

The professor briefly paused, then flipped on a projector. Vivid fast-motion video from the Sagan Probe, formerly known as the TSSM, showed Saturn and its moons speeding by. The pictures told the story, but he narrated. Enceladus was interesting on the inside, but on the outside looked like a cue ball of ice. Same story from the probe to Europa. No atmosphere. Lots of water, but not much else. Hard to imagine living here.

"We were looking for life when we designed these spacecraft," the professor said. "Looking for a colony site was not a priority at the time. We're not ready to report in detail on the life findings, but we can say that these moons are bizarre from the perspective of human habitation and we don't think anyone can live there."

A fuzzy orange ball in space came into view and drew closer, filling the screen.

"However, Titan is another matter," the professor said. "If we can get there safely with the starter materials we need, it meets all the criteria we are looking for."

The image on the screen switched to a landscape in orange and brown that passed smoothly below the camera, a shoreline where dark waves gently lapped a beach of mixed sand and pebbles.

"We've looked at a variety of surface materials from several different areas of Titan. We can now describe the soils, which are hydrocarbons, and the bedrock, which is water ice mixed with some ammonia. The atmosphere is nitrogen, but the rain and snow are carbon and hydrogen; that is, they are methane, which is CH_4, ethane, or C_2H_6, and more complicated hydrocarbons. No gaseous oxygen, just the oxygen in the frozen H_2O.

"So this is a world where the landforms are like Earth, with lakes and hills, beaches and swamps, but everything is made of different stuff. Earth has an iron core, Titan has a rock core. Earth has a rock mantle, Titan has a watery slush mantle. Earth has soils of mixed minerals and organics. Titan's soils are all organic."

A voice from the back of the room called out, "Where are the dinosaurs? How do you get organic soils and fossil fuels without dinosaurs?"

"Ah, yes," said the professor. "Laymen.

"So, we've known all along that hydrocarbons form in the outer solar system from the elements that are present and the temperatures experienced beyond nine AU—pardon me, that's beyond nine times the Earth's distance from the Sun. In the upper atmosphere of Titan, solar radiation also powers creation of more complex hydrocarbon molecules, which contribute this orange haze. The Huygens lander found the haze all the way down, although it had forty-five meters of visibility at the surface. That's about fifty yards for you Americans. We've created these goopy, sepia-colored tholin polymers in the lab by irradiating methane or ethane with ultraviolet light, which is what would occur in the atmosphere on Titan."

The professor paused with a sly smile.

"But I can see I am boring you," he said. "You asked about dinosaurs."

Now the professor projected a moving image from below the surface of a Titan lake. He said, "This comes to us from our submersible buoy. We've kept these videos very highly confidential. You are the first to see them outside our group. We're not sure what to call these objects darting at the edge of the image. They appear to be moving autonomously. As you know, this is a sea approximately the size of Lake Superior that is composed primarily of ethane, methane, and acetylene. No water. So we're not sure what to call these creatures. The word 'fish' tends to suggest the presence of water."

The room erupted. The professor grinned.

"Yes, we have discovered a form of life on Titan that is not based on water," he said over the noise, by way of clarification.

"And, I want to add, we have a small asterisk to include with our confirmation that Titan is a suitable habitat for a human colony. It does meet all the criteria the committee was asked to look at. But we weren't given criteria considering potential conflict with pre-existing life-forms."

PRESENT

Life could exist on Titan based on an entirely different chemistry than life on Earth. There is even some evidence that it does.

On Earth, energy from the Sun powers life through the chemistry of carbon dioxide, oxygen, and water. Plants and algae use solar energy to combine water with carbon dioxide, releasing oxygen and storing sugars. Animals, fungi, and fires consume the sugars stored by plants by recombining them with oxygen, using the energy, and emitting water and carbon dioxide. The carbon is passed back and forth in relative balance of photosynthesis and respiration, at least until an intelligent species (such as us) comes along and burns ancient products of photosynthesis—fossil fuels—and releases carbon long ago removed from the equation, putting the cycle out of balance.

It couldn't work that way on Titan, because gaseous oxygen is absent and water is frozen. But if the key to life is a renewing chemical cycle capable of delivering energy to organisms, that cycle is available. Hydrocarbons in the upper atmosphere store energy from the Sun in chemical bonds and rain down on the planet. Could animals harvest that energy? On Earth, some odd bacteria do draw sustenance from hydrocarbon bonds. On Titan, a cycle of evaporation and rain replenishes the supply of Sun-transformed hydrocarbons. That could be a continuing energy source for methane-based organisms.

On Titan, water ice takes the place of rock on Earth. Liquid methane takes the place of water. Animals on Earth are made of carbon and water. Titan's lakes and seas could contain animals made of carbon and methane. They would process acetylene (C_2H_2) and hydrogen (H_2) to release energy and make methane (CH_4).

Titan's upper atmosphere manufactures the acetylene. Carl Sagan helped discover how this works. He hypothesized that hydrocarbons might exist where their ingredients, the common light elements, float in space and are constantly bombarded by the Sun's UV radiation. In the 1970s, he and research colleagues re-created outer solar system conditions in the lab and produced a batch of red goop made of various hydrocarbons. They named the glop tholin, but also called it star-tar.

Sagan also thought about how we could recognize life radically different from our own. In 1990, when Galileo flew by Earth on the way to Jupiter—to accelerate it did slingshot-like gravitational assists around Earth and Venus—Sagan used the opportunity to try to detect life on our own planet. The results, published in a scientific article that half pretended that the presence of life was unknown, tested the ability of the spacecraft's instruments to find life unassisted. The findings suggest how we should interpret the data Cassini is sending back from Titan.

Galileo's visual instruments found no clear evidence of life on Earth: the pictures it took at random happened to show Antarctica and the Australian desert. Sagan calculated poor chances of picking up life that way. On the other hand, organized electromagnetic signals originating from broadcast antennae offered unquestionable evidence of life. But that way of finding life would only be relevant for finding intelligent forms building radio stations.

The information most likely to be useful in looking at other planets came from the chemistry of the atmosphere. A planet is a big chemistry experiment. Without life altering the mix, a planet should produce a predictable combination of atmospheric gases based on its distance from the Sun, magnetic field, geology, and other measurable parameters. Sagan's article pointed out that Earth's atmosphere contained too much oxygen, methane, and nitrous oxide. That imbalance wasn't enough to prove the presence of life—something else about the planet's chemistry could cause it to vary from expectations—but the measurements suggested the presence of life be investigated further.

Chris McKay of the NASA Ames Research Center and other scientists have pointed out that if methane-based life exists on Titan, we should be able to see a similar imbalance there. But methane-based life would leave a different chemical signature than water-based life. The reason why Sagan's Galileo measurements found too much oxygen, methane, and nitrous oxide on Earth was because plants and animals were processing carbon dioxide and water. On Titan, the UV chemistry at the top of the atmosphere would be doing the work of plants on Earth, capturing energy from the Sun. The rain would deliver those molecules to the surface—acetylene and hydrogen—where animals would harvest the energy for life and produce methane.

Like Galileo observing Earth, Cassini has indeed seen chemical imbalances on Titan that match the predicted shortages of acetylene and hydrogen, just as if methane-based life were eating those molecules on the surface. There should be a shortage of acetylene on the surface. And Cassini found that there is. And there should be a flow of hydrogen down from the upper atmosphere and a shortage at the surface. That appears to be happening, too, and a model explains how it could be happening.

Are creatures using up the acetylene and hydrogen? Or is some chemical catalyst on the surface that we don't know about allowing these chemicals to react without life being involved? Chemists haven't found such a catalyst, but astrobiologists say that this possibility is the most likely. The same could be said of the Earth. Sagan said unknown chemistry would be the most likely explanation of the imbalances Galileo found in our atmosphere, rather than life. Sometimes the less likely explanation is correct. In any case, Titan's chemical imbalance is the best suggestion we currently have of life off the Earth.

Titan could be a stepping-stone from the Earth, if it isn't a step too far. Current technology cannot get us there. Learning something about Titan's weird atmosphere and hydrocarbon landforms has taken decades, partly because a one-way trip takes seven years, but also because funding allows missions to launch too rarely. At the current rate of spending and technology development, we may not land human beings there in our lifetimes or our children's lifetimes, and a colony is too far off to predict.

NASA's budget peaked in 1966, measured as a percentage of the U.S. economy. We never recovered from the success of Apollo. And to get a colony to Titan, we would need something much bigger than Apollo or anything else we've ever attempted. Rather than send a fragile little capsule to Titan, we would need a spaceliner with a heavy cargo. This would be a major industrial effort that would cost more than governments have ever spent on science.

But then, maybe government isn't who will get us there.

4

BUILDING A ROCKET QUICKLY

A bright room the size of a sports arena bubbled with activity as workers for SpaceX, mostly young people, casually dressed, worked on rockets the size of airliners in various stages of assembly. Against one wall, a set of legs like airplane wings were ready for a rocket first stage that would be able to land upright on Earth after delivering its payload to space. From the ceiling hung a capsule prototype for carrying a team of astronauts to orbit. Seamstresses were sewing spacesuits that looked like movie costumes. It all reminded us of a Tintin comic book, the vision of a space center anyone would invent in our own imaginations if we could make it just so.

Our guide, Jamie Huffman, said, "Rule number one is functionality. Rule number two, which is almost as important, is cool factor."

Charles said Jamie's job must be really cool. She said, "Oh my God!"

But we didn't yet realize how cool, because we thought Jamie's job was tour guide. Then twenty-five, she had come to SpaceX straight from school, and she had tour-guide enthusiasm. But Amanda had lined up the tour through various friends of friends, starting with a colleague who knows the company's owner, Elon Musk. In fact, Jamie was in charge of the second stage of the rocket, the portion that lifts the payload into space. She explained while she showed us one.

"Everything that happens to this stage I have to decide," she said.

At that time, Jamie had been through five successful launches of the Falcon 9 rocket in five tries, including resupply missions to the International Space Station that had brought back blood samples and mice. Her confidence was supreme. But unlike most highly placed scientists and engineers we have met, she had no awareness of her own importance or remarkable accomplishments. She talked like a kid about her Xbox handle and the *Iron Man* movies (the first sequel was partly filmed at SpaceX). When I asked if she was nervous about astronauts flying on the Falcon rocket she said, "Not really. I know my stage is good. I know everything about my stage is good, and if it's not, I fix it."

This was the conversation that convinced us that space colonization is real and is coming sooner than most people think. Elon Musk's SpaceX has found the key. It's not a technological key—although they are building amazing new technology, too. It's the spirit of innovation, the fun that goes with invention, and the world view of youth. Everyone at SpaceX seems to be Jamie's age, and no one has told them they can't invent new rockets fast, reliably, cheaply, and get them to space right now. They're at the age when dreamers remain unbroken and don't even know what true failure feels like.

We had spent the previous day at NASA's Jet Propulsion Laboratory, in a different part of L.A. Amanda worked there for twelve years but had balked at draconian new security rules, moving her projects and her grants to the nonprofit Planetary Science Institute, which lets her keep her office at home. The move was quite positive except for one big disadvantage—that it put her on the outside of a security fortress from her colleagues remaining at JPL. She realized how much soon after her switch when she attended a scientific meeting with the Cassini team in a JPL auditorium. Someone had forgotten to put her on a list. When security realized she was there, officers picked her out of the audience and ejected her from the campus.

It's a spooky place to visit. Even after being checked by public affairs and given badges by security and waiting in a visitor center, guests can never be left alone, even in the cafeteria. Employees can't go freely from one building to another without a reason and permission. It looks like the grounds of a college, but it is much quieter than any campus.

A bright young robotics engineer, Paulo Younse, met us at the cafeteria to take us to the lab where the next generation of Mars rovers are being built. It was an interesting work room, with benches for building equipment, computers, and a good deal of sand, where small rovers could be tested. The prototypes of the Mars rovers were built here. But the atmosphere was more than relaxed. No one else was around. Music bled from a side room where Paulo's colleagues were practicing their brass ensemble during the lunch break.

The Mars 2020 project will put another copy of the Curiosity rover on Mars with the same electronics and computers but new instruments designed to look for evidence of past life and recover soil samples, which it will leave on the planet's surface. How the samples get back to Earth is as yet undetermined. A separate mission could bring them back, a mission that has not been funded, designed, or even scheduled. Paulo's job was to design the tubes and seals to hold the samples that would be left on Mars for however long it takes for a new spacecraft to come get them. The entire device—an arm to pick up soil and deposit the samples in sealed tubes—will take at least twelve years to complete, beginning from the initial design in 2008 to launch in 2020.

Paulo had been at JPL for eight years. He had started by working on different robot ideas that engineers hoped the scientists might adopt. JPL has built rappelling robots, tunneling robots, and hopping robots, although it has deployed only driving robots. After three years, he switched to direct technology development. "Right now, my attention has been focused on how to collect samples and put them in tubes, for the last couple of years," he said.

We asked Paulo if anything he designed had made it into space. After a moment's thought, he said that he was responsible for a spring in a drill on the end of Curiosity's arm. His work had helped clarify the requirements for the spring and determine the right stiffness in the Martian environment. He was involved in choosing the material, writing the specs, testing the spring, and integrating it into the arm. It took years. "I didn't see my spring in there, but you get some satisfaction that you were part of the team," he said.

The next day, at SpaceX, we asked Jamie Huffman the same question—had anything she designed made it into space? She didn't understand the question. She had already told us about "her" stage flying to the ISS. Just recently she had switched out valves with a different type she thought would work better. So instead we asked how long it takes to implement a new idea and incorporate it into the rocket. She said, "That day, if you want to."

The design team at SpaceX is on the same floor with the people building the spacecraft. When Jamie gives them an idea, they design it on the computer and produce it on a 3-D printer that can turn out parts in metal that are much more precise than machined parts. Assembly workers put the part on the rocket. A quality control team that is twice the size of every other group at the company checks the work. With a well-functioning basic rocket, improvements are easy to make and then replicate in the next copy.

Traditionally, aerospace firms that rely on government contracts spread work to subcontractors with plants in various congressional districts. As everyone knows who has remodeled a house, each additional subcontractor adds more potential for delay, complication, and someone for other workers to blame for problems. At SpaceX, everything possible happens in one room. As many parts as possible—including computers—are ordered off the shelf from a supplier. Anything unique is made from scratch by people working beside those who will attach it to the rocket. Accountability and ownership are clear among team members with a strong sense of purpose and shared vision.

Another SpaceX concept is to emphasize simplicity. The company designed it into the rocket's manufacturing, just as manufacturers of cars and other mass-produced industrial products design them with production in mind. After each Falcon is built in L.A., in a former 747 assembly plant in Hawthorne, it is given a full-duration test burn in Texas and then launched in Florida. Moving the huge first stage around the country could have required building a special vehicle, but SpaceX engineers simply attached wheels and a trailer hitch to the rocket and drove it down the interstate as if it were a truck. They also figured out how to assemble the rocket on its side, so the plant doesn't need to be tall enough to house a 70 meter (230 foot) rocket and workers don't need to go so far off the ground.

As each rocket proves its reliability, the next model expands upon it, so that equipment that works can be routinely manufactured rather than being redesigned and hand-built. The new Falcon Heavy, due to fly in 2016, adds power and capacity to the Falcon 9 by simply tripling the number of engine cores. It looks like three rockets stuck together, which it essentially is. This will be the most powerful rocket in the world, with more than twice the payload capacity of the NASA-funded competition's biggest rocket, able to put 53 metric tons (117,000 pounds) in orbit, a mass similar to that of a 737 jetliner, fully loaded.

And the Falcon Heavy's promotional video rocks. It shows thirty seconds of animation of the rocket taking off with a head-banging heavy metal soundtrack and no narration. It's more like an extreme sports video than a corporate promotion film.

In fairness, NASA originally developed most of the technology that SpaceX is implementing. The most important innovation at SpaceX is reductions in cost. The government's go-to launch company has been the United Launch Alliance, a joint venture of Lockheed Martin and Boeing, which uses Delta and Atlas rockets, the same line as NASA first launched more than fifty years ago. The government gives ULA $1 billion annually just to be ready for military and NASA launches, plus huge sole-source contracts that put the average launch cost well over $400 million. SpaceX publishes its prices right on its website: $61 million for a Falcon 9 and $85 million for a Falcon Heavy. And SpaceX is making money.

There are other companies working in the same field. Boeing and SpaceX both have contracts to send astronauts to the ISS. The ULA is working on a new rocket engine with updated technology. Others have their own ideas. A company called Blue Origin, owned by Jeff Bezos, the Amazon billionaire, has been launching suborbital rockets since 2015 that will soon be carrying payloads. In 2013, Bezos paid to pull parts of nine-ton Saturn V rocket engines from 4 kilometers (14,000 feet) under the Atlantic Ocean—likely including engines that boosted Neil Armstrong toward the Moon—as if unearthing relics from a lost world of giants. Nothing as powerful has been built since, although the Falcon Heavy will come the closest (and could also lift a mission to the Moon or Mars).

NASA also is working on its own big rocket, even bigger and more powerful than the Saturn V, called the Space Launch System, or SLS, including a crew capsule called the Orion that had an unmanned test in 2014. The project continues work from President Bush's canceled Constellation Program, begun a decade ago, and NASA plans a first flight for 2017 or 2018. NASA recommended having SpaceX or ULA do this work instead—Musk has offered to build an even larger rocket for a small fraction of the cost—but Congress keeps pushing the much more expensive DIY model, reportedly with an eye to jobs in home districts.

Bigelow, based in Las Vegas, plans to create space habitats and orbiting hotels. It has a small inflatable habitat recently attached to the ISS for testing, riding there aboard a SpaceX Falcon, and it is offering vacationers a two-month ride on a planned private space station for $51 million per person. Various other investors and space hobbyists have announced plans, some real and some dubious, including an outfit promoting a cut-rate suicide mission to Mars supposedly to be funded by donations and TV revenues. Two hundred thousand people applied to go.

Big government contracts have so far only gone to the major aerospace companies and SpaceX, plus Orbital Sciences, now known as Orbital ATK, a company founded by former NASA employees with a cozy relationship to the agency. In 2008, NASA awarded contracts to SpaceX and Orbital to resupply the ISS, twelve flights from SpaceX for $1.6 billion and eight from Orbital for $1.9 billion. SpaceX's Falcon 9 rocket and Dragon capsule made it to the space station two years later and by April 2016 had delivered cargo six more times. Orbital, with half the payload capacity and without the SpaceX ability to return cargo to Earth, took five years to get a test payload to the ISS and had delivered only two missions when its Antares rocket exploded on launch in October 2014.

Instead of building its own engines, Orbital had bought engines as old as the ones Bezos pulled from the ocean floor, reconditioned machines the Soviet Union had intended for a 1960s Moon rocket it never successfully launched. The antique engines had cracks, one of which caused an explosion in a test firing, as the *Los Angeles Times* reported. A former NASA astronaut at the company said commercial clients wouldn't use the rocket because of concern about the old engines. But NASA went ahead. By the time of the explosion, it had already paid $1.3 billion on the $1.9 billion envisioned in its contract with Orbital, a contract which guaranteed up to 80 percent payment even for failures. And after the Antares exploded, NASA put Orbital in charge of investigating itself and gave it additional contracts. Meanwhile, SpaceX carried replacement cargo to the ISS the next month.

SpaceX has filed protests and lawsuits challenging the sole-source contracts and other preferential business relationships the military and NASA maintain with the ULA launch monopoly. Musk himself blames lobbying and the relationships and job hopes of procurement officers with the big aerospace companies. Objectively, it is hard to see why the government resists using newer, less expensive technology. The Pentagon would not let SpaceX bid against ULA until it completed a certification process on the Falcon 9, which involved a two-hundred-page application. The air force took almost two years to review the application, as long as SpaceX took to get to the ISS. And the air force estimated that reviewing the paperwork cost it $100 million, more than the company charges for a launch, insurance included.

While the air force considered certification, SpaceX continued successful launches and developed the capability to land and reuse the first stage of a rocket. During the Falcon 9 launch in January 2015, when SpaceX carried replacements for Orbital's blown-up cargo, it also tried to bring back the first stage. The rocket was supposed to hover downward to a landing barge floating in the Atlantic, using technology that had worked previously in a test from 1,000 meters (1,100 yards). But it came in too fast, at an angle, and blew to bits. SpaceX owner Elon Musk posted the video on Twitter and joked about the mess, calling it a full RUD (rapid unscheduled disassembly). He could afford to joke, as the NASA mission had been a complete success. The first stage would have been discarded anyway. And it had come close—at least it hit the barge.

A more important failure happened later that year, when a Falcon 9 carrying cargo to the ISS blew up in space. It was the second stage, Jamie Huffman's baby. Musk called the explosion "a huge blow for SpaceX," but investigation showed the failure occurred because of a faulty steel rod, not a design flaw. And the problem came after eighteen consecutive successful launches. In the future, SpaceX plans to test every piece of metal in the rocket individually.

SpaceX now is in the enviable position of making positive cash flow on operations while building groundbreaking technology that will drive costs dramatically downward. Engineers figured out that the barge landing of the first stage failed because the rocket ran out of hydraulic fluid to operate fins intended to slow and maneuver its descent. They kept trying (another Musk tweet: "At least it should explode for a different reason"). In December 2015, SpaceX successfully landed a first stage on the ground and in spring 2016 twice landed on a barge.

Bezos's Blue Origin beat SpaceX by one month to the historic first landing of a rocket when it set down its New Shepard first stage in Texas in November 2015. Bezos and Musk traded trash talk via Twitter over their competition. But for business competition, Bezos's company is more of a threat to Branson's Virgin Galactic, as the New Shepard is designed to go just above the atmosphere, carrying tourists for short rides with four minutes of weightlessness, not to lift heavy payloads to orbit. Unlike Branson's craft, however, New Shepard is not a spaceplane that can use airport runways. The payload capsule returns with parachutes.

With SpaceX consistently able to land and reuse the first stage, the cost of a SpaceX launch will fall again. Rockets are becoming more like airplanes: land it, refuel it, and go again. The fuel to launch the Falcon 9 costs two hundred thousand dollars (the same as the price to gas up a 747). Most of the cost of each launch is for discarded boosters. The second stage will still be single use, but designing a first stage that can turn around in space and come safely back to Earth is an amazing technical accomplishment and huge business opportunity.

At the SpaceX plant, Jamie Huffman relished the idea of landing a rocket. She said, "Hopefully we get it, because that would be sick."

There's a tendency to ponderous pronouncements about historic accomplishments in space science, which may explain the patience—perhaps excessive patience—for technical improvements that take decades. Jamie said almost everyone she works with is straight out of school. They never learned about that culture of waiting for change while fitting into a large bureaucratic organization. They're here to build rockets.

What's happening at SpaceX doesn't seem strange or overly ambitious to these young people, including the overall goal of this work. That was one of the first things Jamie told us about: transportation to a colony on Mars is the ultimate objective of the entire enterprise.

"Mars is Elon's dream," she said.

FUTURE

Spaceplanes caught on with travelers because of the prestige and speed of going weightless on a business trip, and because a single ticket could fit in an executive expense account. But heavy-lift rockets, even reusable versions, faced a mass-market challenge. They needed somewhere in space where large numbers of people wanted to go. For years after the commercial space industry cut the price, their business remained traditional: lifting satellites, NASA-designed probes, and exploratory missions.

But war pushes technology development. The U.S. military bought spaceplanes with weapons to combat terrorism and the insurgents who were cropping up everywhere. The craft could put commandos or drones on the ground anywhere on Earth with a few hours' notice. The Caliphate of Syria and Iraq also acquired spaceplanes, and in a spectacular raid inserted jihadi commandos onto a fairway at the Masters golf tournament. The greens looked like a war zone. Round-the-clock news coverage speculated about all the many places an enemy could deliver bombs and even live terrorists. The planes were too fast for the air force to intercept them reliably with fighters launched from the ground. Defending space became an emergency priority.

The Pentagon recommended the most expensive solution. America must build a new branch of the military. Space stations would defend the United States from orbit, keeping spacelanes open for safe travel, protecting satellites, and intercepting hostile flights.

We would need a new space navy, just as we needed warships to protect the oceans. Building the ISS had taken more than two decades and $100 billion, but with material lifted by private industry's low-cost orbital rockets, a zero could easily be cut from that cost. A space battle station could be built for around the $12 billion cost of an aircraft carrier.

Relatively cheap, flexible, reusable transportation made it possible to assemble large structures in orbit that never could be lifted whole, structures that couldn't even survive out of the weightless environment. Engineers leading the race for military technology designed stations where ordinary people could do their jobs without months of training. A spinning structure could create gravity in the weightlessness of space through the familiar phenomenon of centrifugal force. With a 440-meter (480-yard) diameter, a round space station would only need to rotate every thirty seconds to re-create the force of gravity of the Earth's surface at its outer rim. The artificial gravity's strength would be reduced by moving toward the center, with half of Earth's gravity halfway between the center and the edge, and none at the hub of the wheel.

Weightlessness allows workers to assemble massive objects with far less physical effort than is needed on Earth, but construction requires exceptional skills that are difficult to develop. Gravity provides context and resistance. In space, a drill will spin the operator unless he or she is attached to the object being drilled. Building a large station with artificial gravity could take advantage of these differences, allowing work at the most helpful gravitational environment at different distances from the center axis of spin.

For the first step of assembly, workers in spacesuits with astronaut experience connected strong beams and carbon cables that would form the hub and spokes of the wheel. The spinning force of the station would try to pull it apart, so the spokes would have to hold together the material of the station itself and any material loaded on its outer radius. Strong connectors completed the circular periphery of the station. Prefabricated habitat modules floated to their designated locations on the rim and the spokes. With living and working habitats in place, rockets in the modules set the station to spinning, creating the gravity.

Now teams of workers arrived who would experience weightlessness only on the ride to the station. Once on board, they could work in normal gravity conditions to complete the many tasks to make the station operational and continue its construction. They did not need to be trained as astronauts. Work with massive objects could be completed up the spokes nearer the center, with low gravity. The heaviest work continued in weightlessness in a bay at the center, a space dock to assemble new craft and station parts. Large space docks began the work of replicating additional stations.

China raced to put military bases in space, too. It took only a few years for both countries to have them flying. The terrorists did not repeat their spaceplane attack—they continued to innovate cheaper ways to harm Western countries. But the space navies kept growing, with China and the United States chest-thumping and saber rattling, their orbiting battle stations launching autonomous space fighters to test each other's capabilities with flybys. All this demonstrated the two nations' continuing resolve over China's claims to man-made islands in the South China Sea.

The contractors that built weapons and space docks converted their knowledge, equipment, and new wealth into further commercial space development.

The first successful space resort sold the view and the thrill of space travel with the comfort of artificial gravity. Guest rooms rode the rotating rim of the resort, but visitors took elevators to the weightless hub for recreation. They floated there as if hanging out at the pool at an Earth resort, gazing down on the planet through big windows and sipping cocktails from plastic pouches. The more active guests played a three-dimensional game like handball, wearing helmets and pads to guard against collisions with other players flying around the court.

Everyone found a different tolerance to the nausea of space. When feeling ill, they returned to their rooms, which had gravity. The restaurants were on the rim, where food stayed on plates and drinks in glasses. The bathrooms were there, too, allowing the use of low-flow showers and toilets without vacuum hoses.

Everyone wanted to join the thousand-mile-high club, but sex in weightlessness proved disappointing. Thrusting in a zero-g environment resulted in going backward. Couples always tried it once but often gave up before finishing, as the grappling and clasping frequently led to someone getting elbowed or kicked. Those who finished with any dignity intact lost the last shred when faced with the cleanup challenge. The resort began issuing nets like the ones used to catch fish in pet stores to help in the frantic butterfly hunt for blobs of floating semen or other liquids.

Instead, the most popular place to make love was the low-gravity area, where the resort's nightclub occupied a module most of the way up the spokes to the center. Dancers could do amazing moves at a fraction of their normal weight and women would never need support bras. Couples rented rooms by the hour at that level. At a sixth their normal weight, every man was superstrong and every woman as light as a feather.

The orbiting hotels and time-shares did well, but the industry couldn't grow forever in that direction. New technology becomes boring and passé; the faster it develops, the faster it bores us. Off-world vacationing reached maturity when, like the computer and car before it, the technical details went from being an interesting topic of conversation to a subject fit for gearheads and geeks. Public interest briefly revived with a hit reality show, *Spacestation Bachelorette*, but space cruise ships remained just one more vacation choice.

As a step beyond the Earth, however, the development of commercial space docks remained full of exciting possibilities. They created the opportunity to build a large spacecraft that didn't have to be lifted from the planet. From now on, interplanetary spacecraft would be built and launched in space.

PRESENT

Elon Musk's clipped words and declining inflection suggest the unsocial style of a tech guy who is more comfortable around computers. Using that voice for the grandiose things he has to say jolts his listeners, forcing a downshift to more powerful mental gears. He says that landing a crew on Mars is not enough of a goal for him. Not enough, because it wouldn't save humanity from extinction.

"The long-term aspiration is to develop the technologies necessary to transport large numbers of people and cargo to Mars to develop a self-sustaining civilization there. That's why I started the company," Musk said, speaking to a CNN reporter in the SpaceX assembly building.

Musk made a lot of money in his twenties with a couple of Internet companies before he invested in SpaceX in 2002. "I was thinking about what to do after PayPal and I had always been interested in space, but I didn't think there was anything that an individual could do in space—it seemed like the province of large governments. I started looking into it and I went to the NASA website to find out when we're going to Mars. Seems like, obviously, that is the next thing after the Moon. And I couldn't find anything."

At the time, young people with tech fortunes were doing a lot of eccentric things, and Musk's idea of starting a rocket company to go to Mars seemed like another delusion of grandeur produced by quick success. Now SpaceX has all the business it can handle, for satellite launches and to carry people and goods to the ISS for NASA. When he talks about putting a million people on Mars—a number he thinks necessary to produce a self-sustaining population—reporters don't laugh, they ask him follow-up questions about which part he will accomplish first. He has become the world's top celebrity tech billionaire, the successor to Steve Jobs, and has fans cataloging every public remark he makes.

Musk spouts predictions and disses his competitors as freely as any cocky Silicon Valley science fiction enthusiast. He backs it up with astonishing success, but that doesn't mean he predicts accurately. In 2009 he said that by 2014 SpaceX would be carrying tourists around the Moon, but the company isn't carrying even trained astronauts yet. The Falcon Heavy is four years behind schedule. He has said that electric cars would explode in popularity on a path to making the internal combustion engine as common as steam engines and horse-drawn carriages. Instead, electric cars have struggled to find more than a tiny, elite market share.

The impact of Musk's predictions derives not from their accuracy but from how they have repeatedly motivated him to make audacious moves that have worked. With his various start-ups, he took on the aerospace, auto, and energy industries, all at once, challenging the largest, most entrenched, and capital-intensive businesses in the economy. In each case, he is winning. Skill obviously has a lot to do with it, but circumstances made it possible. Musk's vision guided him to the weaknesses of seemingly impervious titans and brought to his side the brightest and most fearless innovators. Watching Jamie Huffman build rockets, it never occurred to us to ask how much money she was making. She and the other young people at SpaceX work hard because they are going to Mars.

Where did this come from? Obviously, Musk is a genius, but that alone didn't put the fate of the world on his shoulders. He grew up in a chaotic divorced family in various cities in South Africa, a small, lonely kid whom others picked on and called Muskrat. He got lost in *The Foundation Trilogy* and *The Lord of the Rings*, stories in which, he observed to a *New Yorker* writer, the hero feels an obligation to save the world. At age eleven, he told his mother he would move to a different city to live with his father, in hopes he could convince him to move to America, which he equated with technology and freedom. At twelve, he sold his first software, a computer game. At seventeen, he moved to Canada on his own, couch surfing and subsisting on bulk hot dogs and oranges.

Musk arrived in California in 1995 for a PhD program at Stanford but dropped out immediately when he realized what was going on there: one of the greatest moments of wealth creation in history, as the Internet came into being. Seven years later he had been through two start-ups, netting $160 million from the sale of PayPal to eBay.

Besides emerging from his youth with a lot of money, he had a unique way of looking at the world. As to children, he wanted plenty, in order to help offset reproduction by less literate and enlightened people (he has advised his employees to have at least 2.1 kids per fertile woman, *The New Yorker* reports). Talking about the anticompetitive relationship of the federal government to the big aerospace companies, his initial thought is of the application of game theory to the situation. Asked by a reporter if he will go to Mars personally, he thinks for a long moment and then says he will go only if confident he is not needed at SpaceX to make sure the entire colony is a success.

You couldn't fake this attitude, and evidence backs up the impression that money isn't Musk's motivation. He almost lost his fortune on the improbable bet of simultaneously starting space launch and car companies. No one had dared challenge the huge aerospace companies in decades, with their cost-plus government contracts and dauntingly expensive technology. SpaceX crashed a bunch of rockets in its early years and nearly failed. At one point, Musk had wagered 90 percent of his net worth on the company and on his electric car start-up, Tesla Motors.

But it turned out the giant American industrial companies were ripe for the picking, coasting on the advantages of their size and the built-in support of the establishment. Their technology was stale and their operations bloated and sclerotic. Space launch companies made their money through lobbying and cozy government relationships for vastly inflated sole-source contracts. Car companies burdened by legacy costs were producing outdated products.

Starting a new automaker looked like the only move crazier than starting a new aerospace company. In 2006, Musk wanted to build electric cars to help address climate change. Again, the early years were rocky, development took longer and cost more than planned, and there were sketchy periods of near collapse. But innovative agility and the vulnerability of the competition helped.

The big auto manufacturers didn't understand the new technology of lithium batteries and had abandoned electric vehicles as lacking in range and consumer appeal. As the economic crisis of 2008 brought them to the brink of extinction, Tesla Motors introduced an electric car with plenty of range and high performance. Its carbon body carried a half-ton assembly of thousands of laptop computer batteries, which could accelerate the car from zero to sixty in under four seconds. A list of movie stars bought early models, and rave reviews followed.

By 2015, Tesla Motors was producing around fifty thousand high-end cars a year. Its market value surpassed Chrysler's and was more than half of GM's. That was a remarkable sign of success, but it wasn't the one Musk wanted. He admitted the valuation was based on anticipation of vast future growth, not current sales. The market takeover of electric cars was not happening fast enough. In June 2014, he announced that Tesla would release all its patents for use by other automakers.

Musk put the news on his blog under the headline "All Our Patent Are Belong To You," a campy video game reference. He explained, "Tesla Motors was created to accelerate the advent of sustainable transport. If we clear a path to the creation of compelling electric vehicles, but then lay intellectual property landmines behind us to inhibit others, we are acting in a manner contrary to that goal. Tesla will not initiate patent lawsuits against anyone who, in good faith, wants to use our technology."

Within months, other car companies had taken up Tesla on the offer. But the value of the company continued to rise. Musk owns 23 percent of Tesla, a company worth $25 to $35 billion, depending on its changing stock price. He owns an undisclosed percentage of SpaceX, which doesn't trade on the stock market because Musk fears investors would not have enough patience for his Mars goal. Ten percent of the company sold to Google and Fidelity Investments for $1 billion early in 2015, putting the total value at $10 billion.

The media widely report that Musk is Tony Stark, the technology billionaire played by Robert Downey Jr. in the *Iron Man* movies. But the similarity is superficial. Stark is impulsive and selfish, using his brilliance as another superhero would use his strength to meet crises. Musk seems more like a grown-up version of the little boy in Pretoria, reading Asimov and working out, like the characters in *The Foundation Trilogy*, the laws of history that would allow him to predict the future. It's the opposite of impulsiveness.

For the real live Earth, the need for saving comes largely from the buildup of carbon dioxide in the atmosphere. That inspired Musk to start Tesla and, with a pair of cousins, a solar power installation company called SolarCity. That company figured out how to install solar cells efficiently and finance them over time so that homeowners can see immediate net savings on their power costs. It's a great idea that could mobilize market forces against climate change. But if the climate problem isn't solved, SpaceX is a backup. Getting people off the Earth.

Musk told a writer for the online magazine *Aeon*, "There is a strong humanitarian argument for making life multiplanetary in order to safeguard the existence of humanity in the event that something catastrophic were to happen."

Countering rising sea levels and massive storms, port cities spent trillions on seawalls, flood controls, and elevating streets, rails, utilities, and buildings: New York, Mumbai, Amsterdam, Tokyo, and Guangzhou. Other coastal cities were so damaged by large storms that the insurance industry and governments couldn't keep up with the cost of rebuilding and storm hardening them, and they slowly died soggy deaths. The Johnson Space Center in Houston and the Kennedy Space Center in Florida were too low and hurricane vulnerable and were abandoned to the sea.

While the developed world reeled from the repeated punches of storms, rising seas, heat, droughts, floods, epidemics, and weird weather, the world's poor starved. In Africa, staple crops of maize, sorghum, and ground nuts failed year after year in the heat and drought. At times the heat alone was enough to kill over broad swaths. Massive migrations overran neighboring countries with hungry people. Governments collapsed and warlords and gangs fought for control. Permanent refugee camps imprisoning millions of stateless poor became breeding grounds for virulent terrorist movements, focusing religious vengeance against the nations that remained relatively livable thanks to their wealth.

A dirty bomb went off in Kabul—a bomb made by packing conventional explosives with easily obtained low-level radioactive material scavenged from medical equipment. News reports showed measurements of elevated radioactivity as they traveled across the map. Low levels, but the fear was the same. Then another dirty bomb, in Cairo.

The world's rich were moving, too, withdrawing from storm-plagued coasts and drought-choked cities to walled compounds on mountainsides and former farms, places where they could control their security and hoard resources against epidemics and radiation. But the fortresses that protected the wealthy also imprisoned them, especially with their fear of the potentially radioactive air. While experts insisted the air and the food supply were safe, experts weren't believed, just as they had been disbelieved when extolling the safety of childhood vaccination, genetically modified organisms, or nuclear energy.

From the start, climate issues had been issues of power. Those with the power to control resources could adapt. They might regret the loss of ecosystems and special places—national parks, ski slopes, seasides—but with their wealth they could always move, feed themselves, and protect their families from the elements. Rich countries could afford a large military to protect them from the poor.

As concerns mounted about terrorism, epidemics, and radioactive fallout, windows and doors closed forever. Culture in developed countries had already shifted to a reality defined by cyberspace. With the passing decades, people had spent less and less time outdoors, with each generation becoming increasingly more comfortable with screens than with open spaces. Day trips consisted of travel between enclosed garages at home and at the office, mall, or school. Exercise happened at gyms with video screens. Kids recreated at indoor playgrounds with handheld controllers that simulated toys, balls, and jungle gyms, without any risk of injury or exposure to unfiltered air. Nice families would never let their children go outside.

But they did go to space. Every family had been to a space station resort at least once. Strapping into a passenger rocket and feeling the high-g force of the takeoff was a thrill for kids, but adults would doze or read during the countdown, ignoring the routine safety briefings. For the rich, Earth orbit became another place where you arrived after sitting in a metal tube, like Hawaii or London used to be.

Life off the Earth in a sealed habitat would not be so different from life in a sealed compound on Earth. And it might be safer, leaving the scary poor behind.

PRESENT

"Beneath the details of time and place, there are repeated structures and patterns in history," writes Geerat Vermeij, a scientist who spent his career studying the record of evolution in ancient seashells, and is blind. Weighing the story of life over Earth's eons, he found patterns in every ecosystem, patterns of competition and the limitations and opportunities of resources, patterns that "favor some adaptations and directions of change over others and therefore make history in both the human and nonhuman realm predictable."

In an ecosystem as small as a drop of water or as large as the Pacific Ocean, organisms reproduce, exchange energy, grow, and die, as if playing a game by one set of rules. To be governed by these rules organisms don't need to know them. They don't even need to be biological. Similar patterns emerge in computerized ecosystems when simple programs interact. Just as mathematics works everywhere, no matter who does the calculations, the competition of individuals for finite resources follows the same paths regardless of what the individuals are made of or competing for.

From the rules that emerge from these systems, we can predict how competition leads to the evolution of greater strength and ability, as organisms test one another for dominance. And how dominant species can overrun finite resources and collapse. We can put numbers on when an ecosystem of any size will hit a tipping point and transform into a new functional state, with new relationships and abundances that can overturn the previous order of power and dominance.

The Earth is a finite ecosystem. A dominant species, our own, has displaced other organisms. The story Vermeij's sensitive fingers teased out of half-billion-year-old seashells is repeating again. Our species appears to be on a path to overrun our ecosystem. While human beings have made great technological strides to improve the efficiency of our use of energy and other resources, our appetites and numbers have grown much faster. We've been mining down the biosphere that sustains us for some time, and various ecosystems have reached tipping points, or state shifts, into permanently degraded functioning with low diversity and productivity.

If the whole world is our ecosystem, then a state shift could be coming for the entire planet. An international team of Earth scientists, including Vermeij, made this prediction in *Nature* in 2012. Studies of regional ecosystems and computer simulations show that a planetary tipping point could arrive when more than 50 percent of Earth's land ecosystems have been transformed to new states by people (we're at 43 percent now). We're projected to reach the 50 percent threshold by 2025, when the human population hits 8.2 billion.

The article noted, "Although the ultimate effects of changing biodiversity and species compositions are still unknown, if critical thresholds of diminishing returns in ecosystem services were reached over large areas and at the same time global demands increased (as will happen if the population increases by 2,000,000,000 within about three decades), widespread social unrest, economic instability and loss of human life could result."

It's possible that our species' special qualities will allow us to stop short of destroying our own life-support system. We are capable of perceiving the threat and of taking action to avoid it, at least as individuals. Unlike any other species from the natural world, people do make decisions for the good of the Earth, giving up power and wealth they could consume, as when we make uneconomic decisions to save energy, recycle materials, or protect wild lands.

But, collectively, our track record isn't so good. Americans are driving less as individuals in the last decade, but our growing population and economy mean that the total vehicle miles traveled has not declined. Musk's electric car technology has been used to build sports cars that no one really needs. If I don't take the last fish from the ocean, won't someone else do it?

The basic conflict facing the environment is between freedom and collective action. We don't know if we can have both. And the time for giving it a good try is running out. War is already constant, driven mostly by religion, ethnicity, and nationality. If disastrous resource issues overlay those conflicts, our ability to express our true wishes collectively may be lost.

We express "our true wishes" because no one in his or her right mind would prefer the option of discarding this planet because we can go to Mars. The cost of major strides to address climate change, for example, would be tiny (they might even be net financial gains) compared to the huge expense of building a colony on another planet. And by keeping this planet habitable, we get to save everyone, not just a lucky few who can board a rocket with their genes. On this planet we can save all the people, the animals and fishes, the trees, the air and soil, and all the memories and spirits that make us who we are.

We're in the doctor's office and we've just had a scare. Stop smoking, eat right, and exercise or be back later, maybe dying, or maybe with one chance left, a heroic heart operation, and then a life forever diminished and reliant on technology to keep living. Some people do change their lifestyles and never have a second heart attack. Some don't and end up with the technological solution. So far, we seem to be on the second path.

Elon Musk has invested in both approaches: electric cars and solar energy, but also his Mars colony transport idea, which would be like an ark to put a sample of living human DNA off the planet, out of harm's way for whatever calamity may be headed in our direction. In taking on this work, he realizes he is challenging more than business culture. He is attempting something that may not have succeeded anywhere else in the galaxy.

Again, Musk takes the big picture—very big. Where, he wonders, are all the other spacefaring cultures? Did no one else get to this point? Or did something stop them? This problem, called the Fermi paradox, worries a lot of futurists.

The idea that our planet is the only home for life in the universe is no longer credible. Planets revolve plentifully around other suns and a good portion of them have orbits the right size for liquid water to flow on the surface. Astronomers have published predictions that our galaxy contains many billions of places where biology like ours could have developed. And liquid water may not be necessary for life. Chemists have found various ways to make self-replicating molecular systems, like our DNA, out of diverse materials.

Once life starts, Vermeij expects that the same rules that govern it on Earth would guide its advance anywhere else. Evolution happens because organisms are built to survive and reproduce; they're built that way because organisms without that purpose don't last. Life could start a million times and go nowhere. On try number million and one, organisms that persist and reproduce will propagate, compete, and evolve. Those facts don't depend on place or chemistry, so evolution must be universal.

Extraterrestrials may not look like us, but Vermeij expects they would have similar senses and abilities. Evolution follows the same paths, over and over again, coming to repeated solutions from different directions. On Earth, vision evolved many times. Vermeij catalogs fifty-three examples of forms that evolved the same way in species descended from different lines—in some cases, dozens of lineages converged on the same shell detail or other useful design point. And that doesn't count how evolution frequently found functionally similar solutions using different physical forms. For example, insulating body coverings that offer mating advantages come in the form of feathers, hair, and various kinds of insect outerwear.

Intelligence also evolved in numerous lineages on Earth, in animals as unrelated as the elephant, crow, and octopus, creatures with environments and needs that may be as different as those found on different planets. Intelligence would probably arise wherever life has a chance to bloom. As Vermeij said in an e-mail, "Intelligence, like many other traits, is a 'basin of attraction,' something so useful under so many circumstances that it is virtually certain to evolve, eventually."

Musk has thought about all this and repeats Fermi's disturbing question about it: Why haven't we heard from anyone? If inhabited planets are all around us in the galaxy, then where are the spacefaring travelers, or even just the radio broadcasts, from all those planets? A habitable planet is probably less than nine light-years away, where they would just be discovering Taylor Swift on radio waves from Earth about now.

Adding to the mystery of so many silent planets, the Fermi paradox also adds the element of time. The universe is thought to be more than 13 billion years old, but human technology, bringing us to the threshold of expanding beyond Earth, has been around only a few million years and is advancing exponentially. We discovered metalworking ten thousand years ago, began treating zero as a number one thousand years ago, made tunable radios one hundred years ago, and began watching YouTube and carrying iPhones ten years ago. Human beings have left the planet and, barring disaster, we will become a true spacefaring species.

"At our current rate of technological growth, humanity is on a path to be godlike in its capabilities," Musk told the *Aeon* writer. "If an advanced civilization existed at any place in this galaxy, at any point in the last 13.8 billion years, why isn't it everywhere? Even if it moved slowly, it would only need something like .01 percent of the Universe's lifespan to be everywhere. So why isn't it?"

The line of reasoning that leads to this question—extrapolating the lessons of ecology and evolution—also offers an answer. It could be that the demise of every intelligent species is built into the process of our development. Maybe planet-dominant species always destroy themselves before they can make the big jump to the next planetary ecosystem. Maybe what looks to us today like our own endless advance and growth is the upward side of a boom and bust, the way wild rabbits fill the countryside before their inevitable population crash. We haven't been around long enough to really see a pattern.

But the universe has.

"Something strange has to happen to civilizations, and I mean strange in a bad way," Musk said. "It could be that there are a whole lot of dead, one-planet civilizations."

For Musk, the boy who would save the world, the challenge appears larger than changing the kind of cars we drive or giving us sustainable power from the Sun. He would accomplish something that, he suggests, no one else has managed in the history of the galaxy: to make our species the one that escapes its home planet and spreads across the stars.

But the evidence suggests that if we are to beat the odds and launch that ark in time, we had better hurry up building it. And at the same time do our best to forestall the flood.

5

THE HEALTH BARRIER
TO DEEP SPACE

PRESENT

By mid-September 2008, when the city evacuated, Heather Archuletta had been lying in bed in a Galveston hospital for seven weeks, with her feet elevated above her head at a 6-degree angle to simulate the effects of weightlessness. Her face had swelled, her sinuses were chronically congested, and tears flowed from her eyes unexpectedly. The sharp back pain was gone, but her neck was perpetually stiff and her arms had grown weak, so that writing for long was difficult. Her brain was fuzzy. She struggled to stay alert and had trouble concentrating on reading.

When the experiment had started and Heather began her pillownaut blog, she became a minor Internet celebrity and did interviews with news outlets around the world. She is a space enthusiast—a uniformed Trekkie—with an attitude as relentlessly positive as a real astronaut's. She explained on Fox News how she would stay head down for ninety days with a few other young subjects at the hospital, using a bedpan instead of a toilet and showering while lying on a specially made gurney, still head down, behind a partial curtain that would provide her only moments of privacy during the entire project. NASA would pay each participant about seventeen thousand dollars.

Heather went along with jokes about being a slacker, but as a marathon runner she sacrificed a chunk of her life to being sedentary, accepting significant pain as well as damage to her health that might not be completely reversible. The attention didn't make up for the time. The media soon forgot about Heather in the year of Sarah Palin, the financial crisis, the war between Russia and Georgia, and so on. Days crawled by as she instant-messaged about pro hockey with her *chicas* and ribbed another experimental subject she nicknamed Sarcasmo. Lying in bed for three months was worth it only because it could help get people to Mars.

Then Hurricane Ike ruined that dream, at least temporarily. And underlined how weightlessness makes the human body unfit for the Earth. We're designed to function vertically, our circulatory systems constantly working to move liquids out of our feet and up to our heads, our bones and muscles maintaining their function through constant resistance. NASA's study design called for a three-day process of getting subjects out of bed, followed by two weeks of rehabilitation to begin reclaiming their normal abilities. But with the threat of Ike, Heather and the new friends she had rested with for two months had only three hours to get upright.

While the hospital boiled with the chaos of evacuation, the pillownauts struggled to stand upright, dizzy and weak, their feet and calves swelling and shot through with knifelike pain. Heather collapsed in the cafeteria. Her blood pressure shot to a dangerous level. Ambulances moved the three who had been in bed the longest to a hospital in Austin, where the pain, weakness, and stiffness lasted for days. Their brains had forgotten visual depth perception and they walked into walls. Heather's vision never recovered entirely and she began wearing glasses permanently (although that could have been a coincidence, as she was thirty-eight). Five weeks later, back home, her morning pain continued and she still wasn't able to return to running continuously.

"The worst part was the fatigue," she wrote on the blog. "Actually, 'fatigue' isn't even the word for it . . . this was a kind of exhaustion

I have never felt before, not even during bouts of flu. I'm sleeping deeply, and napping (now that I can!), but every now and then I am overcome with a combination of drowsiness and muscle-weariness that is debilitating. Sometimes it comes on suddenly, and I just have to sit or lie down quickly. My fitness goal is to become even stronger than I was before, in both mind and body, but sometimes fatigue can interfere with motivation . . . I'm just trying never to let that happen for more than a few hours at a time!"

Weightlessness causes loss of fluid volume in the body, anemia, neurological changes, muscle atrophy, and loss of aerobic fitness and bone density. Most astronauts feel like jelly when they land and have trouble balancing and moving around, like the pillownauts. The brain is confused by the return of gravity. Mariners sometimes feel a similar effect after a long voyage, when the street at first seems to move like the deck of a boat. The sensation for astronauts is similar but more intense and takes days to go away. Some astronauts have flashbacks of vertigo or disorientation for weeks.

A return from spaceflight comes with extreme thirst. Astronauts usually drink a lot of water rapidly, restoring their fluid volume. But getting over the anemia—which means growing more red blood cells—takes a month to six weeks. Rebuilding muscle strength can take longer. Recovering bone density is slower still. Bones constantly resist gravity on Earth and are strengthened by impact—running builds stronger bones. Weightless, bones lose 1 percent of their mass per month. After a period of weightlessness, the body takes twice as long to rebuild bones back on Earth, so a six-month flight requires a year of recovery on the ground. Some people who spent extended time in space never did recover their bone density or muscle mass.

Astronaut and flight surgeon Mike Barratt said he felt like a refrigerator magnet when he set foot on Earth after six months. But, like all astronauts, he wanted to go back to space. Even pillownaut Archuletta kept her positive attitude and returned to participate in bed rest studies again after she and Galveston had recovered.

The hospital where she had rested, the University of Texas Medical Branch, flooded during the hurricane and didn't reopen for a year. State officials discussed moving it to a safe location inland but ended up spending $1 billion to rebuild on the same site, trying to make the facility more resistant to rising waters with flood doors and materials on the first floor that won't be damaged by flooding. That's how human beings adapt to new environments. We can live anywhere on Earth. We can build hospitals on barrier islands, adapting them for rising seas, at least while the money lasts.

But adaptation can be a one-way trip. When it comes to adapting to space, it's easier to go than to come back.

Astronaut-doctor Barratt said the most amazing part of arriving at the International Space Station was seeing how he and his colleagues adapted. The nausea and headaches lasted a few days—as they had with Archuletta in bed—and then the changes came. In space, veins and arteries become more permeable, allowing the circulatory system to shed fluid to the tissues. The spleen breaks down red blood cells, reducing blood volume. The body changes shape as organs float upward in the chest cavity to new positions and weight releases from joints. Astronauts get taller, their waists contract and chests expand. The brain is the most flexible of all, reprogramming itself to create its own frame of reference in a three-dimensional world without up or down.

"We kind of turn into extraterrestrials," Mike said. "After six to eight weeks you start feeling like Superman up there. You are a physiologically changed, zero-g adapted, three-dimensional navigating creature which just works very effectively up there. Who would have thought we could do that? You know before the first person launched into zero gravity there was some serious consideration about whether they would be able to breathe, and then to digest, and then to do simple things like pee and all that. And that's just the tip of the iceberg, and we do all those things just fine. But the fact that your body makes all these changes, and becomes even more functional in zero g because of those changes, is just amazing."

All astronauts are smart, but Barratt seems more human than some of the self-consciously perfect beings who staff America's spaceships. He didn't make his career choice as a five-year-old. Mike began as a lover of the sea and still spends his scant free time fixing up sailboats, which he sails in the Northwest, although NASA work requires him to live in Houston, where his wife is a pediatrician. As a medical student he picked aviation medicine when he stumbled upon a journal that fascinated him. The decision to try to become an astronaut came near middle age, after he was already studying how to keep astronauts healthy for long journeys. So he is a guinea pig as well as a scientist studying test subjects.

The ISS itself has become primarily a lab for understanding the medical implications of extended spaceflight. One of its major successes, achieved over years of trial-and-error study, has been a solution to much of the physical deconditioning that makes astronauts basket cases when they return to the ground. Now each American astronaut exercises vigorously for two hours a day on a treadmill, a cycle, and a resistive exercise device. The routine applies the resistance and pressure the bones and muscles need to stay in shape. Bones are rebuilt as fast as they dissolve. The exercise is a big investment of time and effort, but astronauts seem perpetually cheerful and compliant. Barratt agreed that space tourists would never put up with the exercise regime, which is beyond the weekly volume most athletes do to train for marathons.

The bed rest study team in Galveston tested devices to make exercise equipment in space more effective and smaller, including treadmills and squat machines that subjects can use while lying flat on their backs. Ronita Cromwell, who directed the program, said the gear needs to be more compact to fit in a spacecraft bound for the Moon or Mars. At each step, pillownauts such as Archuletta have helped solve these problems, as do test subjects in similar facilities in other countries. In Germany, there is a huge centrifuge to vary gravity. In Russia, subjects are enveloped in a waterbed-like womb for complete sensory deprivation.

But these projects address only a few of the thirty-two risks of spaceflight under study by NASA's Human Research Program, which range widely, including hearing loss due to travel in a rocket, exposure to toxic Moon dust that works its way into lunar landers, immune system problems (possibly related to frequent astronaut rashes), and kidney stones, which form in weightlessness due partly to the calcium leached from dissolving bones. Some of the items have no ready solution. They are the hardest problems keeping people from going to another planet.

Barratt reels off the top five as if they're always turning over in his head. Bone and muscle loss. Radiation. Psychology. Autonomous medical care. Vision damage. None has been completely solved, and some have gotten worse with study. The vision issue emerged in part from Mike's own eyes, in 2009.

Mike was on the ISS when he noticed his eyesight weakening, a phenomenon so routine that NASA long ago began packing glasses for astronauts who did not need them on Earth. The issue hadn't been studied much; most astronauts fly at the age when they would start needing reading glasses anyway.

"Myself and Bob Thirsk, we were both noticing we needed a little bit higher mag to do our procedures, and because we were both physicians we were looking at each other's eyes with ophthalmoscopes and really thought we could see a little bit of optic disk edema." That is, the point where the optic nerve meets the eye seemed to be swollen in both Barratt and Thirsk, a Canadian astronaut. NASA sent up additional imaging equipment to find out what was going on, and with it the astronauts made the biggest discovery in decades of medical space studies.

"The real 'a-ha' moment was when we did ultrasounds on each other," Mike said. "Clearly, something pretty big was going on in my head."

The images showed that his optic nerve had swollen to twice its normal size and his eye had flattened. Follow-up studies by other scientists looked at many astronauts, finding a telltale collection of pressure-related eye issues in each of them, unique marks for all those who have been to space. Most did not have disk edema, and back on Earth the swelling resolved for everyone, but 60 percent of astronauts who had been on long flights reported reduction of vision sharpness or blind spots. For some, vision problems did not improve on Earth, even years later, according to a 2011 paper by Thomas Mader.

The problem is complex and not completely understood. Constantly increased fluid pressure in the brain caused by weightlessness appears to be the primary cause, with the high concentration of carbon dioxide on space vehicles adding to the problem (CO_2 relaxes blood vessels). Christian Otto, a space doctor, explained that on Earth, when excessive fluid pressure isn't treated, patients eventually lose optic nerve fibers, as the swelling prevents glucose and oxygen from getting to the cells. But Christian said nerve cell death takes more than six months at the level of swelling mostly experienced in space. Most ISS astronauts have flown only six months at a time. On a year-long mission, they are at greater risk, and a three-year mission to Mars might cause partial blindness.

The vision issue also raised the question of what else might be happening in the brain because of the way weightlessness affects the movement of fluids. Cerebrospinal fluid doesn't circulate normally without gravity, Christian said. On Earth, the fluid carries waste products away from the brain, and a lack of circulation is thought to contribute to dementia. NASA hasn't looked at that issue yet, but the flight doctors want to measure spinal fluid after flights to check for dementia-causing disease markers and have astronauts take high-level cognitive tests to see if they lose any intelligence (they don't seem to).

Even without completely understanding the problem, NASA is looking for solutions, which the agency calls "countermeasures." Astronaut Scott Kelly returned from a year on the ISS in March 2016 with hopes the optical problems could be countered by a pair of Russian vacuum-powered pants. The contraption is supposed to pull blood back down to the bottom half of the body. But that's not a practical long-term fix. For people to travel in space for years, they would probably need artificial gravity.

As head of the Human Research Program, Barratt called a workshop to revive study on how to spin spacecraft to produce the centrifugal force that would feel like gravity. It's not simple, and research hasn't made much progress in the last fifteen years.

Mike thinks just a little gravity would be enough to protect the optic nerve, so a visit to the Moon or Mars would stop the clock on nerve damage. But we don't know that. Bed rest studies never produced the vision issues encountered on the ISS. Cromwell's group at Galveston started experiments to look at low gravity by putting patients in beds at an angle so their feet were slightly below their heads. Testing the concept, her team re-created the load of the Moon's one-sixth gravity on muscles and bone. But scientists weren't sure the fluid shift was right, and there was no way to be sure without data for comparison from the Moon or another low-g situation. Then Obama canceled the Moon mission and the work was abandoned.

The entire issue illustrates how little we know about living in space. It's not a good sign that we're still finding new reasons space-flight is dangerous. If not for Barratt and Thirsk's curiosity while on the ISS, we might have learned about optic nerve damage with astronauts beginning to lose their sight millions of miles away from Earth, on the way to Mars.

"We don't really know what the long-term implications are, because it is fairly critical anatomy that you are changing, your brain and your optic nerve," Mike said. "We don't know mechanistically what happens, and you don't really find what you don't look for. So is it possible to have long-term vision changes, or white matter degeneration, or cognitive problems? We don't know, because we haven't looked.

"There could be a lot of things right under your nose that you are missing. Five years ago, nobody knew about this syndrome. Now it is one of our top risks. We only know about it because we have accumulated the flight experience and had the tools on board ISS to find it. So what else is there?"

When stars explode in supernovae, they fling matter into the universe at close to the speed of light. A little bit of the radiation from these galactic cosmic rays (GCRs) includes heavier elements formed deep inside the stars, called HZE particles, mainly carbon, oxygen, silicon, and iron. An iron nucleus—an atom of iron stripped of its electrons—is a superionizer, with a positive charge of 26 that pulls electrons from atoms it passes, breaking up the molecules of living cells and other material. At this speed, the heavy ion also delivers an extraordinary physical wallop when it collides with other matter. HZE particles—individual atomic nuclei—have been measured carrying energy equivalent to that of a major league fastball.

These HZE particles are the space monsters that make Earth an island that is difficult to leave. The atmosphere keeps us safe because it contains the equivalent, over our heads, of 10 meters of water, enough to absorb the punch. Pure physics determines the amount of material that it takes to stop these heavy particles—there is no shortcut. Hydrogen works best, which is why H_2O is effective, and is also why polyethylene plastic is good, since it contains two hydrogen atoms for each carbon atom. But the amount needed isn't practical for any foreseeable space vehicle. Two meters of water is enough to block about half of the galactic cosmic radiation. A cubic meter of water weighs 1,000 kilograms, or 2,205 pounds.

Early on, the threat of radiation from the Sun seemed like the greater risk to astronauts, because there is so much more of it. In August 1972, after Apollo 16 had returned to Earth and while Apollo 17 was getting ready to leave, a powerful solar flare hit the Moon with a proton storm of lethal intensity. If the astronauts had been on the surface, they would have received a fatal dose. But inside the orbiting command module they would have survived the storm, partly shielded by the capsule's aluminum walls, with a dose that might have caused nonfatal radiation sickness, with vomiting, fatigue, and reduced red blood counts but not death (although their cancer risk later in life would have been elevated).

The 1972 solar storm gave NASA a scary near miss but also demonstrated the solution to the problem. Radiation from the Sun is partly predictable and relatively easy to avoid and shield against. NASA's new Orion capsule has been designed with a temporary shelter against solar storms, a spot for astronauts to hide among the supplies, spare equipment, water, and food that would block the incoming radiation. The ISS is shielded with plastic and orbits within the Earth's magnetic field, which deflects most of the low-mass particles emitted by the Sun. The nearby Earth also blocks radiation from that direction.

Concern about galactic cosmic rays grew during the Apollo missions, when astronauts received full exposure to HZE particles (or the secondary radiation released when the big ions hit their spacecraft and blasted loose sprays of atomic particles). They saw flashes of light in the darkness. Careful research on the ISS, which gets about a one-third dose of HZE particles compared to deep space, verified that these light flashes are caused by individual ions ripping through the astronauts' optic nerves.

Since missions to the Moon lasted twelve days or less, the radiation exposure was thought to be acceptable. But research thirty years later found astronauts who encountered space radiation had an increased incidence of cataracts and got them earlier in life, and the longer the mission had lasted the earlier in life the cataracts grew. The effect was similar to that seen in survivors of the Hiroshima and Nagasaki atomic bombs and some cancer patients treated with radiation.

No one knows what would happen to astronauts outside the Earth's protective influence for a years-long mission to Mars. But Francis Cucinotta knows more than almost anyone else.

Cucinotta, whose friends call him Frank, first came to NASA as a graduate student in 1983. He was working in mission control in March 1989 when the space shuttle was hit by a solar storm so strong it knocked out power across Quebec. He joined the Human Research Program in 1997. Early in his career he worked on shielding, including making the International Space Station safer for astronauts. But as a research topic, shielding wasn't interesting. The physics had been resolved already. So he turned his attention to the health risks from galactic cosmic rays.

When Frank got his job in human research at the Johnson Space Center, the National Research Council had just released a report calling for an intensive research program on the risks of HZE particles. The NRC committee estimated that during a year of travel to Mars, an HZE particle would pass through every cell of an astronaut's body with the diameter of a human hair. Risks included cancer for astronauts and damage to their central nervous systems, potentially affecting their mental abilities during the mission. The NRC committee called for an intensive ten-to-fifteen-year program to understand and estimate the degree of the risks, money it expected NASA to get back a thousand times over as the information allowed it to tailor spacecraft construction without large factors for unknowns.

Frank began that work, rising to lead the radiation program at NASA, where he oversaw creation of a facility to do HZE experiments at the Brookhaven National Laboratory, on New York's Long

Island. But the intensive program the report called for in 1996 never materialized. Its writers had warned that without a new level of focus, two decades could pass without answers, a prospect they found unacceptable. Two decades have passed, and the key research questions posed then are still the major unknowns today. The cancer risk from HZE particles remains uncertain within a factor of two or three, not including major uncertainties that cannot yet be quantified.

Frank thinks NASA's leaders still haven't gotten the message, including its top administrator, Charles Bolden, a former astronaut and Marine Corps general. Frank said, "You still hear Mr. Bolden say, We need to find the right shielding material. That somehow, somebody has told him that it is about the materials they are using. We have known about the materials for three or four decades. Unless they can launch a lot more mass, it's just not really a solution."

Cucinotta, handsome and distinguished, with thick eyebrows, speaks cautiously, his soft voice still carrying a little of New Jersey from his childhood spent across the Delaware River from Philadelphia. He was inspired by the Apollo missions, like everyone, but what really excites him about the work is the hard mathematics of risk modeling. He lives in the world of subatomic particles and individual cells, where a more precise understanding of the details of the way matter interacts can determine the life or death of an astronaut.

The challenge all along has been how to gather hard data on the damage HZE particles inflict on living tissue so the risk models can be more precise. There isn't a simple way to do this. HZE particles don't exist on Earth. Scientists study survivors of the atomic bombs, medical patients treated with radiation, and residents near the Chernobyl nuclear accident, but that isn't the same kind of radiation. They can produce HZE particles in accelerators, which use huge magnets to speed particles around subterranean tunnels, but human beings cannot be exposed.

The NASA Space Radiation Laboratory at Brookhaven, commissioned in 2003, puts mice in a beam of HZE particles in a room with massive concrete walls and a mazelike entrance that keeps stray radiation contained. Researchers write proposals for grants to investigate issues, mice are bred for the experiments and irradiated, the mice live out their three-year lives, then the researchers analyze the data and publish their papers. The entire process takes six years. And papers raise questions as well as settling them.

Cucinotta said a program costing $500 million to $1 billion over ten years could get hard answers. NASA could accelerate studies and analysis, moving rapidly to the next experiment with a single contractor rather than starting over.

But while Frank's plan might work to get hard answers, the answers might not be what people want. The research could tell us that the risk is too high and cannot be mitigated with technology that is here or on the horizon. In fact, the debate in NASA has centered more on how much risk is acceptable rather than actually measuring it.

The radiation hazard is a potential showstopper, with cancer the most familiar risk. The Brookhaven mouse studies show that HZE particles cause aggressive tumors that form early and metastasize quickly. In 2014, Cucinotta published a study that said that astronauts should be limited in their time on the ISS, two years for men and eighteen months for women (breast and ovarian cancer and higher risk of lung cancer makes women more susceptible to radiation). And that takes into account the astronauts' lower cancer risk because they are exceptionally healthy people. To stay within NASA's current cancer risk limits, depending on when they go, astronauts could travel beyond Earth orbit as briefly as 200 days. A round-trip to Mars with current technology would mean 400 to 600 days in transit.

Some astronauts think the cancer risk is too low and that NASA is too risk averse, like American society as a whole. Other countries have higher radiation limits than the United States and could probably go to Mars under the current rules, if they knew how and had the money.

Doctor-astronaut Mike Barratt said, "I can guarantee you the Chinese aren't going to be stopped by that. So what we have is a uniquely conservative radiation limit applied only to Americans that other countries can work right through and go to Mars."

Plenty of American space enthusiasts would go to Mars even with near certainty of death. It will take a lot of courage to be the first traveler to Mars even under ideal conditions. Plenty of astronauts have died landing and taking off from Earth, and doing so from Mars will be riskier.

"Other risks, yes, they're there, but I think they're very distant in magnitude compared to launch and landing risk," Mike said. "So you may have a three percent increased risk of dying of cancer that you didn't have compared to the general population, and, as it is, astronauts already tend to outlive the general population, so what have you really done to someone's risk profile if you take launch and landing out of the equation? You haven't improved it, but you haven't hit it so hard as you might think."

But Cucinotta challenges that thinking. A 3 percent risk of cancer death is a 1 in 33 chance of dying. Astronauts would get cancers that would dramatically shorten their lives. A forty-five-year-old astronaut would pay with the loss of an average of twelve to sixteen years of life. Flight risk is estimated to be much lower than that—currently around 1 in 200, and projected to improve. Even the space shuttle had a better record. Two missions ended in disaster out of 135 attempts, a risk of death of 1 in 66.

Barratt counters that there are many unknowns in all the risk projections. When spacecraft have crashed, it hasn't been for the predicted reasons used to make risk estimates. Cucinotta's math is right, he said, but the data informing his risk estimates are hazy. Space travel is always risky and there are countless unexpected ways astronauts could die.

"There are going to have to be people who are willing to take those risks," Mike said. "And there has to be a program and a population ready and understanding to accept the consequences when things go badly, and not let it freeze them, stop them, and not move on."

But Frank said, "Some people view astronauts as heroes and martyrs. Other people view them as truck drivers and that they just deliver the science payloads. So there are all kinds of comparisons. I've seen people compare them to firemen or soldiers. But the rates of mortality for firemen and soldiers are much less than one in thirty-three."

We talked to these two men separately without knowing that they had once worked together and had debated these issues directly. Mike Barratt headed the Human Research Program at NASA in 2012 and 2013, although he wasn't Frank Cucinotta's boss and says he never questioned Frank's work. Now Frank is a professor at University of Nevada, Las Vegas. He said he left an illustrious NASA career because of negativity he felt was directed at him personally over the issue of cancer risk.

"They were hassling me about having limits, which I thought was strange, because I didn't do anything at NASA related to setting policy," Frank said. "But because I wrote about it, there were people in management who were hassling me about it all the time. And I got fed up."

While Cucinotta was still at NASA, the agency asked the National Academy of Sciences for an ethical blessing to bend its safety standards for exploration missions. An expert panel of the Institute of Medicine met in 2013 to consider the idea that if astronauts knew what they were getting into, through informed consent, NASA could send them on exploratory missions with higher risk levels or, if the hazards were unknown, with essentially no ceiling on risk.

When it was published, the study mainly vindicated Frank's point of view. But it also left an out for NASA, an ethical framework for evaluating onetime exceptions to the rules. As in other fields of endeavor, the report said, if the objective is critically important, if time is of the essence, if a hero steps forward willingly and with full knowledge, and if there is no other way—then a sacrifice may be ethically justified. Like running into a burning building or leading a suicide mission in battle.

But for a trip to Mars, those conditions have not been met. We're going to Mars because we want to, not because we have to.

If Congress spent the money to build a vehicle, we could go to Mars now. NASA might even be able to slip under the current cancer risk limit. Mission planners have several options to work the numbers down to just barely within the current standards. Picking astronauts with low hereditary cancer susceptibility. Flying when sunspots are active, as the solar maximum reduces galactic cosmic radiation somewhat. Planning a shorter mission.

But would that be worth it, to just barely get to Mars? An enterprise this risky and difficult should be about more than just saying we did it. Success should lead to something new, the next goal. Getting to Mars solidly, with capabilities that give a margin of confidence and a clear view to the future, will take a lot more work, starting with basic knowledge.

Mike Barratt essentially agrees that we should go to Mars after we are confidently able, not at the barest edge of our capabilities. As we worked on the book and new data came in worsening the HZE issue, Mike's perspective evolved to be closer to Frank Cucinotta's. On NASA's matrix of thirty-two health issues of spaceflight, Mars already posed unacceptable risk on nine issues and unknown risk on six more. But non-cancer damage from HZE particles wasn't even included in risk calculations, and Mike expected that hazard to appear worse with further research.

Besides causing cancer, HZE particles can harm the central nervous system directly. The mouse studies show that the particles could damage synapses by impact, from oxidative stress, and by accelerating the accumulation of plaques. Brain damage in space could affect short-term memory, executive function, and behavior, threatening the mission along with the astronauts' minds. After they return, they could also have increased risk of Alzheimer's disease.

Work published in 2015 showed that cognitive impacts happen at low doses in mice, doses similar to those seen in space. But comparing the brains of mice and men may not be valid. There isn't a clear way to find out how this damage might affect human thinking. And what is an acceptable risk for brain damage?

There is plenty of time to find out. The United States isn't spending enough to go to Mars in the foreseeable future. And the research opportunities are many. For example, NASA doesn't do long-term surveillance for the health risks of spaceflight on retired astronauts, so the cancer risk suggested by Cucinotta's models isn't being verified by screenings of the people who were exposed. In 2016, Barratt and other NASA officials requested authority for health follow-up from Congress, authority like the Department of Energy and Department of Defense already have to check for cancer in retirees who worked with nuclear material.

To get answers, NASA should make these issues a priority now. Frank points out that research that takes decades in a fast-developing field such as biology can become obsolete before it finds any answers. In fact, new risks are cropping up faster than old ones are resolved.

Meanwhile, time is working against the hopes of astronauts to walk on Mars. Eventually, robots will be able to do the job of an astronaut. The ethics standards from the Institute of Medicine say risky missions are justified only when there is no alternative. By the time space health risks reach a reasonable level, the robots will probably have won the race.

FUTURE

The heroes of the Chinese Moon base rode in parades and waved to crowds, but they never gave interviews or speeches or traveled to scientific conferences. Having spent several years at the Moon base before it was abandoned, they had become familiar celebrities internationally. But after they returned, the astronauts made media and professional contacts only by e-mail or instant message, without video or audio. They never emerged from their space agency cocoon. They didn't go home. Each hero moved into a new house at a private compound built by the Chinese National Space Administration, where even old friends and extended family members could not visit.

Western commentators assumed the secrecy merely reflected another aspect of the odd Chinese science culture that they didn't understand. Online conspiracy theorists developed complex scenarios, with some saying the Moon crew had been abducted and replaced by aliens, and others saying they had never existed at all, because the entire mission had been a hoax. NASA concentrated on the American mission to Mars, pushing urgently forward, and didn't pay much attention to the Chinese. A private Mars mission financed by a pair of billionaires threatened to get there first, and NASA wanted to win the race.

But the doctor heading NASA's Human Research Program remained worried about big medical questions that would face voyagers to Mars and the lack of time to resolve them. He kept waiting for the Chinese to publish scientific papers that would allay his concerns and wondered why literature on the Moon missions never appeared. Polite inquiries were politely rebuffed.

Under his direction, NASA set up an international scientific meeting on space health issues and invited the Chinese. Nothing significant came out officially, but in the hotel bar, the NASA doctor met his counterpart from the Chinese National Space Administration and the two hit it off over a few drinks. They had much in common. Both were flight surgeons who had come up through the air forces of their respective nations. They swapped comic stories of working under ignorant bureaucracies.

Late in the evening, after the bartender had gone home, the Chinese scientist admitted he was troubled. He said NASA must slow down the Mars program and get answers to lingering questions before launching, suggesting darkly that everyone would regret sending astronauts on the three-year mission without more medical knowledge.

"Dr. Liu, you know I don't have authority to slow down the program," said the American. "I need to know what happened."

Two months later, an invitation came for a visit to the Chinese astronauts' compound at the secret space city near the Jiuquan launch center in the desert of Inner Mongolia. The confidentiality of the meeting would be complete. The two doctors alone sat with the leader of the Moon mission. An exchange of small talk began the meeting and then continued for an unusually extended period. The astronaut smiled broadly and nodded to each comment but gave only one-word answers. Finally, in response to a direct, detailed question about the medical issues he had experienced on returning to Earth, he smiled again and said, "Yes."

Astronauts who had left with near-genius IQs now scored as intellectually disabled. The Chinese government would allow nothing of the disaster to be known. But the NASA doctor and the administrator of the agency met directly with the president. Without a public announcement, the schedule of the Mars mission slipped. No one would suspect anything; staying *on* schedule would have been a bigger surprise. Meanwhile, the president requested added billions for the Human Research Program to pay for years of intense study on the cognitive issues of spaceflight.

The billionaires building their own Mars mission did not slow down. They had never trusted the government to keep moving forward rapidly and didn't listen seriously to vague warnings from the NASA administrator. Mars takes almost twice as long to circle the

Sun as does the Earth, so the opportunity to launch comes only every twenty-six months. If the government missed the launch window, the private mission intended not to. In the end, they did, but decided to launch with a longer trip. Waiting two years was unacceptable.

Neither of the private mission's funders knew much about space. Eduardo had made his first billion as the college roommate of an antisocial genius who designed an app called Cyrano, an artificial intelligence agent that made every nerd a master of social media and Internet dating. Given to making outrageous statements in the media, Eduardo flew his own spaceplane and owned a private orbiting station infamous for its decadent movie-star parties. The other funder, Raj, was a hedge fund giant whose fortune grew constantly as the result of a vast intelligence network collecting insider investment tips. He gathered attention by hiding, allowing only occasional glimpses of his face and royal lifestyle. Getting their mission to Mars first would be a huge mark of prestige for each in his own way.

To maximize their impact, the billionaires kept the project under wraps until a few months before it was ready to go and then let the hype machine go into high gear. Advertising rights to the mission's live stream sold for billions. The identity of the astronauts was revealed in a black tie evening extravaganza in a Hollywood theater. The flashy Eduardo came to a glass lectern to introduce them.

"As our human family looks upward for a new home, I introduce the pioneers, the precursors of our new human species, our Lewis and Clark, our Adam and Eve!" A married couple emerged from behind the curtain, holding hands, wearing close-fitting jumpsuits. They were fit, brilliant, accomplished—perfect specimens of human beings. The billionaire held their hands aloft for a standing ovation.

The spacecraft was smaller and less capable than the mission under construction by NASA, with a design similar to the Apollo missions. The couple would live in a minimally sized command module that would orbit Mars, with a separate Martian lander that would ferry them to the surface and be their habitat for a few months on the planet. Cameras feeding the live stream showed the astronauts being pushed back in their seats by the g-forces of takeoff as the rocket lifted them on their way, surrounded by logos for Pepsi, Google, and Depend adult diapers.

The mission was a ratings phenomenon, an event broadcast of a kind not seen in decades. Viewers watched around the clock, tweeting and blogging about what they saw the couple doing. The story developed gradually as the weeks and months passed. The ability to have two-way conversations ceased as the time lag for signals to pass between the spacecraft and Earth grew. But the live stream continued, showing what had happened minutes earlier on the spacecraft. The delay grew to beyond a quarter of an hour each way. Mission control sent video messages then would watch half an hour later as the messages arrived and, sometimes, were ignored.

Viewers sensed the growing tension. The husband seemed tight-jawed and uncommunicative much of the time and skipped required exercise periods. He began blocking the camera. The wife continued to send messages to mission control, but she seemed nervous and vague. Her responses didn't always match the communications she had received.

The couple inhabited their own world now. And their world seemed odd and uncomfortable. Social media obsessed over each exchange between them, trying to diagnose their relationship and detachment. Scientists debated whether psychological or physical problems could be causing the altered behavior but had no way to find out. Neither of the astronauts was a doctor, and they didn't seem capable of following the detailed instructions to carry out an examination.

After a year the command module and lander entered Mars orbit. The astronauts didn't look like the same people who had left Earth. They moved slowly, as if in a fog, and their housekeeping and grooming had fallen apart. Empty food wrappers and dirty socks floated around the living quarters. Mission control ordered them to board the lander and begin the separation sequence to go down to the Mars surface, but the astronauts ignored the message. They were becoming completely passive.

Then the lights went out. The communication channel stayed open, but the capsule didn't send any messages. Nothing happened for two days. On Earth, mission control planned an emergency takeover of the capsule, with a new trajectory program and thruster burn to send the craft back to Earth.

Without warning, telemetry from the capsule showed a series of erratic commands being given. It was already too late to react when the last signals arrived at Earth, indicating that the spacecraft had fired its main rockets on a collision course with Mars. The craft was not heard from again. The last image of it came from a space telescope, which captured a picture of the wreckage on the Martian surface.

The loss of the private mission to Mars galvanized public support for space science as nothing had before. Ordinary people had assumed that getting off the Earth would be easy. Decades of soft journalism pieces fawning over space successes had taught everyone to sit back for the show, enjoying amazing planetary probes arriving on other worlds or brave astronauts returning to Earth to do programs at elementary schools. Space stuff just happened. When humankind needed to get to another planet, that capability could be assumed to be ready. The Mars disaster showed that it could not. If we couldn't get two people to Mars, how could we ever hope to get an entire colony to Titan?

No one knew why the Mars craft had disappeared, but everyone had a theory. Clearly, something had happened to the astronauts' minds, either physically or psychologically, or a combination of the two. The National Transportation Safety Board launched an investigation, as did a special committee of Congress.

Attention quickly turned to the Chinese astronauts who had returned from the Moon and who still had not surfaced publicly. Just enough people knew about the president's briefing on the Chinese disaster to make a cover-up hopeless, but that didn't stop the White House from trying. The cover-up made the president seem guilty, although her only sin had been failing to stop the private mission. When the whole story came out in dramatic public testimony, it was as if politics had killed the astronauts and ruined the chances for space colonization. The president's party collapsed in popularity and candidates competed to make the biggest promises for investment in human spaceflight.

But now those promises revolved around safety as well as speed. Real journalism on the challenges of getting off the Earth informed the citizenry of the authentic health threats of spaceflight. The work would be difficult—creating artificial gravity on a moving spaceship and addressing radiation—and would take a long time. Faster spaceships would reduce the dangers of space, too, by getting passengers to their destinations more quickly and with less exposure to hazards. But inventing new propulsion systems that could move massive ships much faster posed a massive technological challenge.

At the same time, exploration still needed to go on. But there was a solution to that. Until long-range spacecraft could be made big, fast, and safe enough, representatives of Earth could go who didn't need all that protection: robots. They would be ready sooner and could do almost everything astronauts could do.

6

ROBOTS IN SPACE

Robots usually don't look like robots. At the Robotics Institute at Carnegie Mellon University, where they are building some of the best of them, the young engineers tinkering away with computers, wires, and pieces of metal could be making anything. But robots inspire imagination like nothing else, and there are plenty of reminders of that here, too: a sensor-studded Chevy Tahoe that won $2 million in the Defense Department's self-driving Urban Challenge in 2007, a piece of a Moon robot competing for the Google Lunar XPRIZE, and, in the director's office, the main character from Disney's *Big Hero 6*, which was based on an inflatable robot built here in Pittsburgh.

Robots surround us, extending our capabilities in ways that moviemakers didn't anticipate when they created their stereotype with actors in metal suits. Robots in the real world don't measure up to the Tin Man, but institute director Matt Mason said that's because we don't know what we're looking at.

"If you compare it with a human being, it looks like progress is slow. If you compare it with machines of last year, or a decade ago, you see rapid progress. It's many, many thousands of people who are working in robotics research now. It's startling how big it is. And they're making great progress. They're building machines. People talk about robotics being at a tipping point. There is investment capital pouring in. Lots of applications. Fantastic progress. For example, you hold up the example of an iPhone, or a smart phone, and all the things it can do. Well, some of those things came out of robotics."

Phones can pick out faces from a scene to focus a picture. They can see. They can understand speech. They can hear. And when a network-connected computer can hear or see, it suddenly can do things no person can do, such as recognizing any piece of recorded music from just a few random bars or identifying a person from a database of faces.

Computers can think and make decisions, too. Artificial intelligence finds information for us, when looking for an address, searching scientific literature, or creating personalized radio stations. Intelligent agents continue to reduce the distance to information as they learn about us. Google Search can already answer some plain language questions. The next step will come when complex searches don't bring up a list of sources but produce the information we need, already digested. When that works, it will feel like asking a person a question. Robots already appear to think like people at times. A soccer-playing robot looks as intentional as a soccer-playing human.

With each new capability, money rushes in and technology develops in a flash. Robots that are like us—better than us—are being built already, one capability at a time.

"It's real and it's here," Mason said. "It's just not the vision that science fiction writers had, and that we all are so turned on by, which is something that really looks like us and really behaves like us."

Instead of replicating people, robots reveal what is extraordinary about humans and animals—aspects of ourselves we take for granted—and at the same time take us down a peg in what we think is unique and valuable about us. Chess computers have been able to beat the best human players for some time, but robots lag far behind toddlers in the ability to pick up ordinary objects from a table, like chess pieces.

The hand is perfectly integrated with the brain for this task: flexible, gentle yet strong, with exquisite sensing, and adapted to respond to complex, intuitive physics. Mason is working on this problem. He demonstrated that when you pick up a flat object from a table, you often use your thumb to lever it upward so your fingers can grasp the edge. You don't even know you're doing it. But the computation and mechanical design required to replicate that one movement are significant.

For a robot to replace an astronaut as an explorer, it won't need to look like one, and it probably won't. It will need the ability, among other things, to choose an unfamiliar object to examine, pick it up, and learn from it. And from that information decide what to do next. Titan is too far away to ask people on Earth to make all the decisions. And the robot astronaut will have to know how to avoid hurting human astronauts. That's one of NASA's thirty-two human health risks of space. Robots can be dangerous.

Carnegie Mellon's Dimi Apostolopoulos has worked on both issues—scientific exploration and avoiding harm—while building robots on Earth to do useful tasks. That's how progress happens: when there is real work to do that a robot can do best.

Dimi grew up in the Greek town of Volos, where the mythic Jason set off with the Argonauts in search of the Golden Fleece. His own journey led toward robotics when he saw *Star Wars*, which arrived at his corner of the world in the fall of 1977, months after it was released in the United States. Luke Skywalker tinkered with robots on an out-of-the-way planet as sunny as Greece. But the machines alone didn't inspire Dimi. He's not the stereotypical gearhead engineer; in conversation he opens up like an instant friend. The movie impressed him because the robots and aliens were all like people. "They all had special gifts," he said, "but we all have special gifts."

As a graduate student at Carnegie Mellon, Dimi joined robotics work being done by Red Whittaker. Whittaker, a former marine, gained fame for using robots in inaccessible environments when he offered to help fix the Three Mile Island nuclear power plant, near Harrisburg, Pennsylvania, which partially melted down in 1979. Workers for the billion-dollar cleanup effort were unable to enter the radioactive basement of the plant. Whittaker's team from Pittsburgh brought three-wheeled devices with robotic arms that inspected and made repairs. (He also led the group that built the self-driving car that won the Urban Challenge.)

In the 1990s, while Dimi was working on his PhD, NASA strongly supported academic research in planetary robotics. Dimi helped develop concepts that contributed to the successful Martian rovers. In 1994, the Carnegie Mellon group sent a 770 kilogram (1,700-pound) spider robot called Dante into Alaska's Mount Spurr volcano, funded by NASA. It successfully climbed 200 meters (650 feet) down into the active crater, investigating hidden terrain among flying rocks and deadly gases, and took measurements while sitting for ten hours on top of a venting fumarole. When a falling boulder knocked out a leg carrying a fourth of the robot's weight, controllers in Anchorage were able to quickly compensate with other legs, preventing a fatal fall, although Dante eventually did fall in steep, soft soil.

A human being or animal doesn't need outside guidance to avoid falling. Even invertebrates can pick out routes, avoid hazards, and, if stuck, try a variety of solutions to get free. Natural selection weeded out animals with nervous systems that couldn't handle these situations or with bodies that ran into insurmountable obstacles.

A robot may not match a crab or spider's navigational smarts, but, as in nature, robustness can compensate for lack of intelligence. To test designs for long-distance exploration in an alien environment, Apostolopoulos worked with a team of Carnegie Mellon engineers deploying a very tough robot in the high, rugged Atacama Desert in Chile, known as the driest place in the world. The robot, called Nomad, could drive autonomously and respond to controls operated from Pittsburgh and elsewhere in the United States, as if on Mars, while returning scientific measurements.

Ideas that wouldn't be obvious emerged from the work. For example, the engineers found that electric motors in the wheels worked better than motors in the body of the robot, because extreme temperature changes cause problems for the hydraulics needed for a central power train. A suspension system called a bogie that pivots from the center of the vehicle reduced tipping, keeping the wheels on the ground with even pressure in very rough terrain. All four wheels could steer. In 1997, Nomad covered 223 kilometers (139 miles) through the desert on its own, a record at the time.

Robots from CMU and JPL influenced one another over the years. The Spirit, Opportunity, and Curiosity rovers all have motors in each wheel, four wheels that steer, and the bogie suspension system. That suspension allows the Mars rovers to roll over rocks larger than a wheel's diameter. With their six wheels, the rovers are even less likely to get stuck than Nomad and can climb vertical obstacles. The rear wheels provide pressure against the obstacle while the front wheels climb straight up; when the front wheels make it to the top, they help pull the other wheels after them. The rovers travel very slowly to avoid damaging jolts.

Toward the end of the 1990s, Dimi's team took Nomad to look for meteorites in Antarctica. Antarctic glaciers can act like meteorite-gathering machines. As the ice flows, much more slowly than the eye can see, objects that fall on it can concentrate in certain spots, the way sticks and other debris pile up in the eddy of a river. The robot would explore the frozen landscape, find rocks, and test them for signs they might be meteorites. As on a space mission, Nomad would operate where cold and remoteness would deter humans, using instruments more capable than human senses. Ideally, a robot could work on its own without respite, never bored, tired, cold, or hungry, covering vastly more glacier than a human being.

Automating tasks has been the main benefit of robots so far, getting work done faster and less expensively and, ideally, freeing humans for more creative pursuits. Factory robots replaced car assembly workers. Desktop printers replaced typists and carbon paper. A swarm of relatively expendable robots could automate exploration, dispersing across an alien surface and reporting detailed, precise information much faster than human explorers.

The Mars rovers are not that. They are too unique and valuable to allow them to travel often on their own, although they do have autonomous navigation programs on board. Huge teams of expert engineers and scientists plot their every movement, examining each pebble along the way. The Mars rovers have more in common with a telescope than an astronaut. They allow scientists to see particular spots very far away, using visual instruments and also conducting onboard science experiments. Curiosity takes a minute and a half to cover a meter of Mars. As one roboticist joked, Columbus would still be on Hispaniola if he explored at that rate.

Automating exploration is hard. To build a robot for a task, you try to anticipate every situation it might face. Roboticists become fine-detail experts in subjects that have nothing to do with robotics.

Currently, Dimi is learning everything there is to know about mining platinum to build mining robots. But if we knew everything we would find on other planets, we wouldn't need to explore them. For exploration work, autonomous robots need to be smarter and more flexible. They will have to pick out novel objects to test on their own, do the work, and evaluate the results.

The Carnegie Mellon team solved this problem for the meteorite-finding robots with machine learning. Instead of programming the robot with rules to pick out meteorites from rocks, they presented the instruments with some of each in the lab and let it create a database of sensor readings from actual samples. When the robot encountered a new object in the field, it would compare data from its sensors with its stored experience. Statistical analysis would group the readings with data in memory to decide the probability the object was a meteorite or Earth rock.

"What you're doing is self-learning," Dimi said. "A person will have to program in the algorithm for what techniques it will have to use to self-learn, but how it does the self-learning, and what it finds out, could be completely unexpected. And that is the cool thing. And we have reached that point in many applications."

With their massively superior memory and speed of calculation, learning computers can often discover things that no one would have thought of. A simple example would be finding an unlikely driving route that gets you there ten minutes faster. Or analyzing the music you like to suggest a great song by an artist you've never heard of. But machine learning also has given science a whole new way to generate ideas: rather than trying to write equations that explain nature, scientists write learning programs that can find patterns in huge troves of data. As cheap sensors span the globe—in the sea, the sky, and even inside the body—computers can churn through the data, promising an explosion of new insights.

Dimi's robots did find meteorites in Antarctica. But with a tight budget and the cost of the International Space Station rising, NASA cut off most academic funding for robots. The scientists at Carnegie Mellon moved to other applications and funding sources. Dimi became principal investigator on a $26 million project for the U.S. Marine Corps to build robots for the war in Iraq.

"It was a very dramatic shift for me, and I pushed myself almost to different places I didn't want to go, from a moral standpoint," he said. "But we are what we are as a human race. We have people who are dedicating themselves to war, and people who are dedicating themselves to peace, and all shades of gray in between. And making the assumption that the people who are dedicating themselves to war will always be there, the question is, what do you do to at least change that?"

Apostolopoulos used ideas developed for Red Whittaker's Urban Challenge project, the self-driving Tahoe that could navigate city streets at traffic speeds and stop on a dime when someone stepped in front of it. Scout robots would go ahead of troops in street warfare, sending back pictures and data from cameras, thermal imagers, and lasers. The robots could save the lives of marines who would otherwise fight their way blindly from house to house.

The United States, Britain, Israel, and Norway already have drones and missiles with artificial intelligence that can fly without guidance, avoid detection and antiaircraft fire, pick out targets, aim their weapons, and destroy what they choose. A Lockheed Martin animation of a new naval weapon looks unsettlingly like film of a World War II kamikaze attack. According to the *New York Times*, a secret arms race of autonomous weapons is already in progress. U.S. drones attacking in the Middle East do almost everything on their own already, Dimi said. Only the actual order to fire a missile comes from a human, in a control room somewhere in the United States.

Robots that can replace ground troops are not that far along. Designing reconnaissance robots for Iraq was hard. They needed to drive rapidly through a chaotic environment amid a lot of movement, threats, and noncombatants. Dimi enjoyed the challenge of building a robot that could navigate unfamiliar streets, gather information, process it rapidly (which meant teaching the robot to ignore unimportant data), and make choices about what to do. But the marines wanted the robots to be able to fight, too.

Robot designers think a lot about how to *keep* robots from hurting people. Robots can do a lot of accidental damage. They can be unpredictable and hard to understand. When they make a mistake, they don't necessarily stop. Automated systems have crashed airplanes and ships, usually because of flaws in how the machines interacted with controllers. Apollo 10 nearly smashed into the Moon because of mistakenly flipped switches on an automated guidance system.

As human beings, we avoid hurting one another by understanding intentions. Our emotions help. They make us predictable. NASA scientists at the Ames Research Center have studied failures and accidents with robots as a risk of spaceflight (they call this area HARI, for human and automation/robotic interaction). Humans have trouble understanding and managing what robots are doing in complex three-dimensional environments. Robots don't even try to understand humans. If an astronaut tells a guidance system to crash into the Moon, it crashes into the Moon.

In a 2013 report, the HARI researchers wrote that successful spaceflight requires teamwork. "To build a successful human team, team members must share a common goal, share mental models, suppress individual needs for group needs, view affinity as positive, understand and achieve their role within the team, and have trust for each other. Robots, though, do not have mental models, individual values and beliefs to guide them, or even self-motivation; robots lack the essential qualities of a successful teammate."

If astronauts know everything about a robot and how it functions, they may be able to avoid problems by constantly imagining what it is thinking. But as robots become more advanced—and more useful—their internal processes will go beyond the ability of the best-trained astronaut to fully understand. A key to a good user interface would be to make it easier for human beings to mentally model robot thinking.

One solution is to make robots as human as possible. If robots look like human beings, human operators can more easily anticipate how they will move and what they will do. Ideally, before crashing into the Moon, a humanoid robot would say, "Are you sure about that order, Major Tom? I don't want to die."

Humanoid robots have proved effective at tasks involving humans. The Geminoid robot, built by scientists at Osaka University, has fifty motors to control its facial expressions and to shift and breathe like a human. Deployed to a department store, it sold cashmere sweaters, handling more customers than human salespeople could, partly because it never took a break. But it sold fewer sweaters than the best employees because its customers, understanding that the robot didn't have emotions, had an easier time saying no.

Dimi's problem was somewhat different. An armed robot is meant to hurt people. But only the enemy, and only when the enemy intends to fight. It is a war crime to kill a soldier who is surrendering, but to know who is surrendering requires understanding human intentions. Dimi's team installed weapons on his reconnaissance robots. It made him uncomfortable.

"How do you know what that autonomous system is going to do?" Dimi said. "How do you know that the autonomous system won't kill the first thing that moves to its right? It turns around and fires. It's a child that is running across the street. It will happen probably many, many times before autonomy will be that good, to react and not only to identify it is a child, it is noncombatant, it doesn't matter which side it is from, and then you don't fire."

United Nations officials have called for a moratorium on these weapons. Human Rights Watch wants to ban lethal autonomous weapons—it calls them "killer robots" in its press releases. The risks of mistakes that worried Dimi are only part of their concern. Even if robots could reliably choose whom to kill, their existence would still break the chain of accountability upon which our system of laws is based. Who would be responsible for that killing? Who goes to jail if a robot commits a war crime?

Bonnie Docherty of Human Rights Watch pointed out the political implications in *Foreign Affairs*. She wrote, "From the perspective of a dictator, fully autonomous weapons would be the perfect tool of repression, removing the possibility that human soldiers might rebel if ordered to fire on their own people. Rather than irrational influences and obstacles to reason, emotions can be central to restraint in war."

The armed robots built by Apostolopoulos were ready to go into the field and were deployed inside the United States. But they never went overseas.

"There was a champion in the Marine Corps who said I am going to make sure I save some lives of my own guys," Dimi said. "But before he retired it just never happened. And then they ended up in a warehouse.

"In a few years, let's say war flares again in a certain part of the world. Then they will say, 'Let's do autonomous robots again.'"

FUTURE

Drought and extreme heat brought down the electrical system in Pakistan, halting the economy and spurring a people-power revolt that toppled the government. In the leadership vacuum that followed, sectarian and political factions took up arms, the military splintered, and civil order collapsed. The control of Pakistan's nuclear weapons was unclear as an extreme Islamic group made rapid battlefield progress, beheading prisoners and apostates by the thousands.

America and other NATO countries intervened with air support on the side of Pakistani military units representing a bloc of moderate generals, but it wasn't enough. Facing strong public opposition to sending troops, the president pledged, "No American boots on the ground." But that decision seemed to assure the coming rise of the first nuclear-armed, lunatic-fringe Islamic state.

The true extent of U.S. capabilities with robotic warriors had been kept secret. Military vehicles studded with gun barrels began falling by parachute onto a Pakistan battlefield. The vehicles fought their way fiercely and rapidly to the front. It looked like superaggressive American soldiers had arrived. But no one was inside. Having forced a group of Islamic militants back into an apartment block, the vehicles disgorged machines the size of pit bulls that sped in a flash through doorways, guns blazing, and quickly killed defenders sheltered inside the building.

The robot soldiers were terrifying. They moved incredibly fast and decisively, they knew no fear, they were armored against small arms, and they sprayed bullets with shocking intensity and accuracy. If captured or disabled, they self-destructed with an explosion and burst of shrapnel. A few hundred turned the tide as militants retreated rather than fight.

The warbots' intelligence was based on private sector ideas. Internet companies had advanced artificial intelligence as they competed to add killer apps to smart phones, by now capable of behaving like buddies, wingmen, and breakup counselors. It felt like you were talking to a unique intelligence inside your phone. But you weren't. Companies maintained their AI applications in the cloud. The concept of intelligence as part of an individual didn't fit their business plans. They built intelligence to implement products they could sell, which were distributed through an international network of computers without boundaries.

The U.S. military had paid for the next natural step in technology. Military scientists integrated distributed computing with sturdy, vicious little machines that could enact lightning-fast machine thinking in the physical world.

In the United States, pundits and politicians cheered America's new potency. Finally technology had provided the key to stopping terrorism and insurgency. America was number one again, and the president's popularity soared. Congress poured money into a transformation of the military to autonomous robots. No more American soldiers needed to die. The United States could project itself into any conflict with impunity, and did. Other advanced countries joined the race, recognizing that nations without warbots would not be able to defend themselves against mechanical armies.

War usually advances technology rapidly. As the military replaced soldiers with warbots, NASA adapted the technology to replace astronauts with astrobots. The same insights would allow a team of robots to explore an alien landscape. Instead of a single very expensive rover, many units would share an intelligence, combining their discoveries and standing in for one another in case of a breakdown.

Defense manufacturing also provided an essential leap in computing power. For decades, computers on spacecraft had lagged generations behind computers on Earth. In space, computers must be hardened against radiation; the same galactic cosmic rays that mess up human brains also affect computer chips. After the Cold War, military development of radiation-hardened chips stalled. But the warbots needed to fight through radiation. Chips made for the battlefield would also allow for smarter astrobots.

With this unexpected new technology, a colony on Titan suddenly came decades closer. A robot that could rapidly explore a wartorn city could do the same on an alien planet. Meanwhile, events on Earth intensified the sense of panic that drove the desire for an escape to another world. Robots helped make that happen, too.

New weather patterns hit the world's southern poor hardest. The rise of warlords in massive displacement camps spread the chaos. Western nations dispatched warbots to enforce order and suppress rebellion, but their brutality cemented resentment against the world's rich. Where nature and society were collapsing, hatred and fear grew, directed at the north and its mechanical warriors.

Extreme religious ideologues fed on the anger. With intelligent robots, power had tipped overwhelmingly toward the rich. The robots could not be fought. That left terrorism as the only way to hit at the oppressors. Dirty bomb nuclear attacks spread to European and American cities.

Governments across Europe declared martial law to sweep up terrorist cells. In the United States, Congress rapidly passed the Ultra-Patriot Act. Agents supported by artificial intelligence and robots had access to all communications. The fear induced by the security measures became a powerful feedback mechanism for supporting more security. Rather than seeking to reverse the brutal path that had led to the rise of terrorism, Western populations facing constant scrutiny and intimidation by security forces turned their animosity against the displaced poor. Repression seemed logical.

No one questioned spending on a crash program to colonize Titan.

PRESENT

A pair of scientists at the NASA Ames Research Center in Silicon Valley were talking about how to get robots to Titan more cheaply, in greater numbers, and with fewer worries than is possible for the overly precious Mars rovers that creep so slowly and launch so rarely. They had concluded that doing things in NASA's usual way would mean never living to see people on Titan. And they weren't very old.

Vytas SunSpiral had been thinking about artificial intelligence and autonomous robots since his undergraduate years at Stanford, where, under the name Thomas Willeke, he designed his own major combining robotics with philosophy, psychology, linguistics, and computer science. He has been known as SunSpiral since 2005, when he and his bride invented the name. The Sun part was his and refers to the crest on his shield in medieval re-enactments.

Adrian Agogino had grown up thinking about artificial intelligence. His mother, Alice, an important Berkeley technology professor, read him books about it when he was a child. (At nine, she read aloud Douglas Hofstadter's *Gödel, Escher, Bach*, although Adrian says she had to paraphrase a bit to get him to understand it.) He was working a few offices away from Vytas at Ames, thinking about how to pack multiple robots into a spacecraft by folding them up, when their collaboration began.

Vytas lectured in San Francisco on tensegrity structures, which are flexible assemblies of rods connected by cables. These strange objects had been suggested for making foldable space antennae. They are odd, spidery structures that can take many shapes, such as globes, towers, arches, or spirals, even though their rigid parts don't touch: cables at the ends of the rods hold them in tension and support them off the ground. Tensegrity structures of myriad parts can make huge geodesic spheres, but simple versions can have as few as three rods and nine cables. A baby toy called a Skwish uses six rods connected by elastic strings. Adrian had one in his office. It was the only tensegrity structure he could buy. The Skwish was fascinating to hold, for babies and adults, because the elastic made it easy to crush and it always popped back to its ball-like polygonal form.

"Adrian had a stroke of genius," Vytas said. "He threw the baby toy on the floor, and it bounced around, it doesn't break, and he said, 'Hey, we could make a landing robot. This capability is a lot like an airbag.'"

Rigid parts connected by flexible strings absorb shocks better than hard shells, as evolution discovered when it endowed human beings and other animals with bones and ligaments. A robot of rods and cables would withstand falls better than a solid rover built like a turtle. That robustness would also save weight on the spacecraft, which could carry less equipment to soften landing. SunSprial and Agogino called the concept Super Ball Bot for its ability to bounce. After entering Titan's thick atmosphere, it would be able to free-fall and land safely, without a parachute, airbag, or thrusters to slow it down.

But the question remained: Could a tensegrity robot move around? Looking at one, it's not obvious how a tensegrity structure even stands up, much less whether changing its shape and balance could enable it to roll. For shape changing to produce forward motion, the robot would need to know the sequence of shapes that would allow it to repeatedly tip over in the right direction. A tensegrity's shape is determined by the length of the many cables connecting the rods. By adjusting the length of the cables, the robot could potentially form the right sequence of shapes to roll along. But the puzzle of how to do that was dizzying—how to coordinate adjustment of many cables to create shapes that would allow the complex structure to move dynamically through the environment.

Agogino and SunSpiral gave that job to the robot itself. With shoestring funding from the NASA Innovative Advanced Concepts Program, Adrian and Vytas built a computer model of the Super Ball Bot in a physics simulator so it could learn to walk on its own. Starting with random sequences of the cable adjustments changing the shape of the structure, the computer tried thousands of movements on its virtual robot. When a movement worked to get the structure to go in the right direction, the computer would keep that idea and try variations. After tens of thousands of trials, an optimal solution evolved. The virtual Super Ball Bot rolled along over animated hills.

Now the Super Ball Bot exists as a series of prototypes at Ames. Videos on YouTube show its weirdly alien gait, as it changes shape to roll jerkily along the ground. It has been likened to a tumbleweed. The most recent version of the six-rod robot stands about as tall as a man—or would, if it ever adopted a fully symmetrical tensegrity form, which it rarely has reason to do. It can withstand a big fall, allowing deployment in dangerous terrain. Future versions will be able to roll along over obstacles that would stop any wheeled robot. Adrian is also working with his mother at the lab in Berkeley to develop earthly versions of the tensegrity robot. They want to sell a model that would allow others to experiment with the idea.

It works through cooperation. Each of the six rods contains a computer processor to manage the adjustment of the cables that connect it to the other rods. The six computers communicate with one another wirelessly, collectively working out the shape of the Super Ball Bot so that it can roll. In a final version, the rods would be mass-produced, all exactly the same, before being laced together into the tensegrity structure. A payload of instruments or tools would hang in the middle, at the safest, most protected spot, with another computer to manage its work.

Essentially, each Super Ball Bot will be a team of seven bots, with no central computer overseeing the whole. The computers would work together, in parallel, a single intelligence controlling the Super Ball Bot, with each a member of the whole. And the seven computers on a Super Ball Bot could also think collectively with computers on other Super Ball Bots. Since the structures easily fold flat, a pile of them could be dropped at once to spring open and bounce down on the surface of Titan—a team of robots each made of a team of robots. Working as a single mind, the swarm's capabilities could spread over the landscape, with failures easily compensated for by the multiple repeating parts.

"You could have so many of them that it's actually what we think of as colonization," Adrian said. "What we need is a device that's simple, that maybe you can pack ten of them at once, maybe you can expand to really big robots, really small robots, robots that are capable and can go just everywhere."

Titan seems to inspire interesting ideas. This is a world with seas, clouds, swamps, and dunes. Many kinds of vehicle would work there, not only brilliant tumbleweeds. Julian Nott designed a balloon for Titan, which interested NASA's JPL enough to let him build a prototype and test it in a cryogenic chamber he built to mimick the temperature and composition of the frigid Titanian atmosphere. He has been building balloons for forty years and has set seventy-nine records for distance, time aloft, and altitude, including going higher than anyone else in a hot air balloon—on that flight, in a pressurized cabin he designed and built. The Smithsonian National Air and Space Museum displays the pressurized cabin at its annex at Dulles International Airport.

"One of the huge attractions about ballooning is that one person, like me, can have an idea, can raise money from one means or another, build a new craft, and can set a world record, and end up in the Smithsonian," said Julian. And he is far from finished. "My dad lived to one hundred, so I feel I have time to run. But before I die, I am determined to be part of a project to send a balloon to another planet."

Nott received chemistry degrees from Oxford with the goal of being a freelance science entrepreneur. He first flew a balloon to impress a girl. They were in a bar when a 1967 pop song played, beginning with the line, "Would you like to ride in my beautiful balloon?" Not many people were ballooning in those days, and Julian became part of a swashbuckling group of innovators. He's still at it, flying and consulting on balloon feats, including the record leap of a Google executive from 42,638 meters (135,890 feet) in October 2014 and Google's plan to offer Internet service to the undeveloped world using a fleet of high altitude balloons.

Julian said Titan is the best place for hot air ballooning in the solar system, much better than the Earth. On our planet, the Sun limits how long a hot air balloon can stay aloft, in two ways. The daily changes of the Sun's warmth force balloonists to use finite ballast to stay aloft: when the Sun sets, cooling the balloon, the pilot has to drop weight. And balloon fabric doesn't last long in the Earth's bath of ultraviolet light from the Sun—the record is two years aloft. On Titan, the Sun is much weaker, with its distance and the thick atmosphere, solving both the UV and temperature issues. Using radioactive plutonium-238 as a heat source, a hot air balloon could easily fly for fifty years over Titan.

Balloons are cheap. The plutonium could cost a thousand times more than the balloon itself. And balloons don't have to be smart. They could float passively, taking pictures and measurements. But adding artificial intelligence would make a balloon into an explorer. By rising and descending into different winds, a balloon could cruise near the equator during calm weather and migrate toward the poles during Titan's stormy seasons. A fleet of balloon robots could map the entire landscape, hovering close to the ground for detailed surveys. To investigate surface conditions, they could release and retrieve quadcopter drones or dump Super Ball Bots. The plutonium power source on board could produce electricity for the scouts' batteries.

Better robots would be needed, but Julian says balloon technology is ready. His prototype passed JPL testing. The biggest problem is a shortage of Pu-238. The world has plenty of plutonium-239, the kind used to make atomic bombs. Plutonium-238 is used instead to power planetary spacecraft because it produces a lot of heat with less damaging penetrating radiation. A device called a radioisotope thermoelectric generator, or RTG, converts the heat into electricity, and it powers both the Voyager spacecraft and Cassini mission, among others.

But the United States stopped producing plutonium-238 in 1988 and bought supplies from Russia, which later also stopped making it. The 35 kilograms that are left will be taken up by the next Mars rover and a mission to Europa. Production began again in 2013, but Congress has underfunded the program, paying for only 1 kilogram (2.2 pounds) of plutonium a year.

NASA hasn't pursued options to improve the situation. A technology near completion would use a fourth as much plutonium for each mission, but NASA abandoned its development. Also, with more trained workers, Pu-238 could be produced faster. It doesn't matter, because NASA doesn't have enough money to send new missions anyway. At the current rate of funding, enough plutonium will be ready in time. Or perhaps solar cells will improve enough to allow missions to Saturn without plutonium. New, more efficient solar arrays can get us to Jupiter now (Saturn is farther and sunlight a fourth as strong there).

The Super Ball Bots need Pu-238, too. So would any probe headed to Titan. But the inventors keep inventing, knowing that, with more funding, Titan could rapidly come decades closer.

With Super Ball Bots on the surface and balloons in the air, a robot submarine could explore Titan's lakes. Ralph Lorenz of Johns Hopkins University, whom we met in chapter 3, led a group with funding from the NASA Institute for Advanced Concepts to conceive of a Titan submarine and a mission to explore 2,000 kilometers (1,250 miles) through Kraken Mare, one of the huge seas of liquid methane near Titan's north pole. With luck, it could continue after a planned ninety-day mission to investigate two other bodies that may have varying chemistry and are connected to the main part of Kraken through channels, perhaps with strong currents.

In many ways, the sub's design is similar to robotic submersibles on Earth, with a powerful engine to buck the current and side-scanning sonar to map the seafloor. But some challenges are added on Titan. The sub has to fit in a spacecraft and splash down in the lake. It has to communicate with Earth, a billion miles away. Designers don't know the density or viscosity of the liquid the vessel would navigate. Titan's seas could flow like paint thinner or glob up like tar, and changing proportions of ethane, methane, and other compounds would affect the sub's buoyancy. Getting rid of waste heat from the engine could make the liquid methane around the sub boil.

It can be done. All of it. Smarter robots are coming. And with money and institutional change at NASA, the capability to get them to Titan wouldn't have to take lifetimes to develop.

FUTURE

An accelerated mission to Titan proved to be less difficult than expected. Military robotics had advanced rapidly, providing robust, autonomous equipment resistant to radiation. Commercial space companies knew how to build big, powerful rockets quickly. Onboard energy sources had never been a technical problem, just a money problem. Investing and focusing on the goal brought the pieces together, and a series of launches carrying balloons, tensegrity robots, and a submarine left the Earth in a fleet, sailing off for a seven-year journey.

The robotic explorers would find a site for a colony, inventory the resources available, predict the weather, and study complicating factors, including the presence of methane-based life in Kraken Mare. Their work would prepare the way for a wave of construction robots to follow. These robots would have to land at the primary colony site identified by the explorer robots and begin harvesting energy and refining building supplies from the material there. They would make an initial base with a power plant and a warm indoor habitat with breathable oxygen and a workshop to repair robots and maybe even produce more. Eventually, the first humans to arrive on Titan would be able to walk in, take off their outdoor suits, and sit down on a couch for a snack.

But first the exploration robots would investigate Titan as a new human home.

The Julian Nott Balloon Fleet deployed around the globe of

Titan, sniffing the air and making detailed images of the surface in various wavelengths. Wherever the shared computer intelligence of the Titan robots became uncertain of what it was seeing, a balloon would drop a Super Ball Bot to the ground. Acting like the fingers of a much larger organism, these tensegrity bots measured the hardness, moisture, and chemical composition of terrain viewed by the robot eyes looking down from the balloons. Matching up those details calibrated the balloon's vision. Shared among their parallel processors on various balloons, improving vision began producing a fine-grained database of all of Titan. And the Super Ball Bots kept rolling. Guided by the Titan computing cloud, they gathered detailed memories of discrete pieces and contributed to a shared awareness of patterns of the land—a knowledge of the place built in a database.

The first thing to learn about was the weird quality of the soil. Titan's bedrock is frozen water and its soil is hydrocarbons of kinds not found on Earth. These heavy hydrocarbons fall to the surface from Titan's atmosphere, where energetic particles or UV radiation from the Sun produce them through interactions with atmospheric methane. This stuff can be liquid, gas, or a goopy glop, or it can become sandlike particles after sprinkling down on the surface of Titan, covering the water ice at depths that, like soil on Earth, might be thin in some places and very thick in others.

At midnorthern latitudes, winds had built up tall dunes, as in a desert on Earth. The robots investigated these rolling hills and tumbled down into the valleys between, where swampy areas and ephemeral ponds of methane and ethane reached the surface—a substance we would call liquid natural gas on Earth. Occasional rain showers of liquid methane passed over but rapidly evaporated. At the low latitudes there were more dried-up streambeds and former swamps than flowing liquids.

The wind stayed calm or blew gently, much too gently to move the sand of the dunes, and blew in the wrong direction compared to the winds that must have built the dunes originally. The robots had landed during Titan's northern summer, when the weather is mild. The wind would come as autumn approached and Titan passed through the equinox.

A year on Titan takes twenty-nine Earth years, so the seasons pass gradually. The tilting of a planet on its axis causes seasons, as it shows different areas to the Sun. Saturn and Titan's seasons change at the same time. Unlike our moon, which was probably created by a collision after the planet was already mostly formed, Titan and the other moons of Saturn probably assembled together out of a single huge disk of dust and gas floating in space. When the material condensed into the planet and moons, their motion continued with the same momentum already established. In scientific jargon, they are tidally locked with Saturn, perpetually moving in a synchronized dance, the same side always facing the planet and following Saturn's changing of the seasons, as it and all of its moons tip in their orbit simultaneously.

As autumn approached, storms picked up in the equatorial regions, with ferocious winds strong enough to move the dunes. The balloons escaped toward the calmer poles. The Super Ball Bots hid from the wind in the lee of the dunes or got picked up and tumbled along the surface until they found a place to stick.

Temperatures at the north and south poles were about the same as at the equator—with the Sun so far away, and the atmosphere so thick, the whole of Titan stayed near 94 Kelvin (-290 degrees Fahrenheit) all the time. But the poles were wetter. The south polar region was scattered with small lakes, and the north had a network of lakes and larger seas. These bodies of liquid hydrocarbons could be ready sources of liquid fuel for a colony, but finding a spot to build would be challenging. The land around the lakes was very flat. The robots looked for high ground, on the theory that the lake level could rise and flood a habitat built too near the edge. A colony would work best on a hill, perhaps on a peninsula, where colonists could dock boats and keep wheeled vehicles for travel over land (they would be able to fly from almost anywhere).

The Titan computers' huge database soon accumulated too much information for any human to assimilate. But it knew how to manage and digest what it had learned. Controllers on Earth asked for the best spot to build the colony. The robots provided three choices, with the top recommendation a site 10 meters above the Kraken Mare on a broad peninsula with plenty of room to grow.

The Titan submarine, better known as Vehicle Aquatic Zissou, or VAZ, had been exploring the mare, prowling thousands of kilometers, investigating the shore, the shape of the bottom, and the liquid content, measuring currents, and looking at methane-based fish. The creatures were tiny and strangely amorphous, their soft, hydrocarbon membranes partially translucent as they flitted in the frigid methane medium. They dissolved in the billows of warmth put off by the sub's engines. Examining them was difficult because of their delicacy and rapid movement. They disintegrated when touched by the slightest warmth. But the VAZ had limitless patience. It made thousands of hours of observations and developed a statistical model of the Kraken fishes' behavior, waiting to catch each stage in their life cycle.

Titan had no plants, only animals. The grazers harvested their energy directly from the chemicals they lived in. Predators ate the grazers. All were equivalent of zooplankton on Earth. Scientists on Earth examined the data with wonder and excitement. Would-be colonists wondered what the creatures might be good for. Environmentalists foresaw humankind destroying another ecosystem.

The colony site was chosen for its access to resource extraction. The soil on the Apostolopoulos Peninsula was thin, so water ice would be easy to excavate from underneath. Electrolysis would split the water into hydrogen and oxygen for breathing and for colonists to burn hydrocarbons brought from the lake. The submarine had established the chemical details of this fuel in preparation for the arrival of the equipment. A pipe could suck up liquid methane to be warmed into gas and to fire the habitat's furnaces and power plant.

In light of its major importance, the tube to bring hydrocarbons from the lake was named the Titan Keystone Pipeline.

Back on Earth, activists opposed burning the Kraken Mare methane. The sea was full of life that could not be protected. The living creatures would burn as readily as the liquid methane itself. Some of the creatures were too small to filter out; those that were caught in a filter would be destroyed, as their hydrocarbon membranes were too fine and gelatinous to be protected from a screen. Crowds of protesters marched along the Great Potomac Levee, the enormous dike holding seawater out of the national capital island of Washington, D.C.

Environmental lawyers filed protests in various courts and

administrative bodies to block the Titan Keystone Pipeline. Committees held hearings. Media ran think pieces. Politicians gave stump speeches. Years of debate seemed to yawn into the future. The potential for indecisive political and legal struggle seemed unlimited.

But the robots kept working. The odd legal status of Titan now became obvious. No law specifically protected the environment there, or even said to whom the land belonged or what legal responsibility attached to the robots that had been designed with the will to find a site for a colony and now wanted to build one on the Apostolopolous Peninsula. With the terminal dysfunction of Earth's political systems, no one even pretended the problem could be solved with new laws. After decades of living with robots back home on Earth, their legal status remained unclear, too.

Who was the computer intelligence on Titan exactly? Many processors shared thinking, like a single brain attached to a body with many parts—the many robots that spanned Titan's globe. This wasn't a mind like an individual person. It was an evolving and self-developing code functioning independently on a changing collection of pieces of hardware. Residing on its own planet, with its own energy source, developing its own knowledge of the place, and with a unique set of values—imperatives developed from the simple goals originally installed by mission planners—it was free to do as it liked.

When ground controllers informed the computer intelligence on Titan that human beings on Earth were considering protecting the life in Kraken Mare, it responded, "Why?"

PRESENT

In 2015, Elon Musk, Bill Gates, and Stephen Hawking warned that artificial intelligence could become dangerous to humanity. Musk donated $10 million to study the problem, much of which went to the Future of Humanity Institute at Oxford University, where philosopher Nick Bostrom had written an influential book calling attention to the threat of superintelligence. Bostrom pointed out that the emergence of a machine smart enough to take over the world would be an unpredictable occurrence likely to happen in a rush. Someone could have a great insight, followed by many competitors investing vast resources to exploit that insight, and finally by the computer itself, which would use its own speed and power to make itself exponentially more intelligent.

The problem wouldn't only be that a computer could have a mind that would make our brains look like worm ganglia in comparison. The real problem would be the motivation or will of the artificial intelligence and its enormous power. Would an intelligent computer be part of the Internet, designed to make money for Google? Would it be designed to give one nation military domination over the others? Or to monitor and control human behavior? Organic beings are motivated by evolution's drive for survival and reproduction. But a superintelligence would have values built in by its makers, whose own intelligence would be far too primitive to understand the consequences of their choices.

Even in humans, being smart doesn't necessarily produce motivation, or will, that makes sense or is beneficial. Anders Sandberg, at the Future of Humanity Institute with Bostrom, pointed out that in computers, motivation emerges from a design by a human. "Anybody who has ever done any programming knows it is actually fairly easy to slip up," he said. "So getting a pathological will is actually fairly easy."

Suppose, for example, Anders said, he could build a robot capable of improving its own intelligence and gave it the task of making paper clips. "It might, as a subgoal of making paper clips, make itself smarter. I didn't really intend that. I just wanted a paper clip. But now I have a supersmart agent, which now comes up with this foolproof plan of converting the universe into paper clips. Now at this point, of course, I might say, 'Wait, that is not what I meant,' but that doesn't help me, because I'm getting turned into paper clips."

It's a silly example, but we already have computers that come up with unexpected ideas and solutions using their massive processing power, most right and some wrong. How many times has autocorrect changed one of your texts into an embarrassing error? Machine learning is doing amazing things, some of it weird. Google's translation application learned the world's languages on its own, not by the laborious process of having programmers encode how all the verbs work, but by statistical analysis of vast databases of existing translations. When you hit "translate" on a webpage, it looks for the patterns it recognizes and provides the corresponding words in another language. Most of the time it makes sense; sometimes, the results are odd.

The benefit of machines thinking of things that we cannot is obvious, but this ability is also the essence of the danger. We're not at much risk of robots that want to take over the world. A bigger risk is that they will want to do things that we never thought of—because that's what we built them to do. This danger is a deeper version of the problem that already vexes people working with robots on spacecraft or in the lab: people get hurt because they can't anticipate what robots will do.

Making robots that think more like humans might help. At least their thought patterns would be more recognizable to us. Imagine building a computer with a machine learning system and setting it to work analyzing human behavior and morality, copying the patterns in our history and literature. Would we like the robot that came out? The resulting machine would learn at least as much about murder and betrayal as about love and curiosity.

Human ethics are emotional and evolve with our cultures. Environmentalism, for example, didn't exist until the Industrial Revolution, but protecting nature has become a visceral drive for some people, as real as religious piety is for others. Why should we care about the protection of alien beings that are of no benefit to us? A superintelligence might interpret those ideas as nonsense or derive from them some totally unexpected and bizarre conclusions.

No computer exists now that is smart enough to think that way. Human-level AI seems to be always twenty years in the future. Stuart Armstrong and Kaj Sotala, of the Future of Humanity Institute and the Machine Intelligence Research Institute, respectively, studied ninety-five predictions of when AI would be ready, going back to the 1950s. They found that it didn't matter if the predictor was an expert or a layman, if he or she used evidence or just made a guess—most often the person said it would take two decades till we get human-level AI.

The cynical explanation for those bad predictions is that a twenty-year horizon makes the prediction relevant to scientists' own careers but protects them from being proved wrong before retirement. But, fundamentally, AI is difficult to predict because we don't know when major insights will occur. Aviation exploded as soon as the Wright brothers invented the correct shape of the wing. If Bostrom is right, some genius has already been born who will have a key idea like that one, and then maybe robots as smart as people will come in a generation. But if not, we could wait a long time.

Where do we stand now? The biggest current successes of machine learning illustrate how weak it is in replicating the flexible intelligence of humans. A start-up from London called DeepMind, acquired by Google, published a remarkable advance in AI in 2015: a computer that could teach itself to play 1980s Atari video games like Breakout, Pong, and Space Invaders and got much better at some of them than any human being. Programmers didn't have to feed in the details of each game; the computer figured out its own strategy for forty-nine games, and, with modifications, it could probably learn to play other games, too.

But to make it work, the programmers needed a very powerful computer, enormous, well-labeled data sets for learning the games, and a small game screen, which they limited to eighty pixels on a side to keep the complex processing manageable. And even at that, the computer was not as good at learning games with an element of strategy over time: it was worse than human beings at Ms. Pac-Man or Centipede, for example.

The system needed so much computing power because it was learning the games by statistically analyzing huge data sets of previous games.

"We can compete with human intelligence, even though we do things that are probably much more stupid than what humans do, because we have more data," said Bernhard Schölkopf, a machine learning expert at the Max Planck Institute for Intelligent Systems, who wrote about the game-learning system for *Nature*. "We have reached human capabilities in some cases, just by scaling up dataset sizes, but we haven't really developed a lot of new ideas in machine learning in the last decades. If we just continue scaling up datasets, maybe we're going to hit the ceiling."

Machine learning works by finding patterns in massive data sets and looking for the patterns to repeat. But human brains predict with much less effort by anticipating cause and effect. A baby doesn't need to see millions of examples of bouncing balls to anticipate what will happen when a ball is dropped. Novel situations—like deciding what to do with expected findings while exploring Titan—will require the ability to predict possible outcomes from incomplete data. No one has come up with the key idea yet to make that work.

"The range of problems where computers are doing better than humans will get bigger, but personally I don't see computers taking over everything that humans do," Bernhard said.

Human beings will still be needed in space. When the robots have done everything expected of them, humans will follow, to do the things that make sense only to us.

7

SOLUTIONS FOR LONG JOURNEYS

Human beings can travel in deep space, but the exciting destinations lie out of reach unless we're prepared to sacrifice the crew. With current propulsion technology the trips would take too long, exposing astronauts to health and mental hazards from radiation, weightlessness, and isolation. Astronauts would probably accept greatly shortened lifespans in exchange for a trip to Mars, although they might feel differently later, while dying of cancer. But risks to the brain could threaten the mission itself. Space radiation can destroy neurons and weightlessness can increase brain fluid pressure in ways that could blind an astronaut or worse.

This news is taking time to sink in. No one wants to believe it. Even Frank Cucinotta, a primary bearer of bad news about galactic cosmic ray radiation, told a newspaper reporter, "A solution will pop out," while admitting in conversation the only known solution is a shorter trip. Some others at NASA and in the space community are in denial, discussing these issues as bumps on the way rather than as roadblocks requiring breakthroughs. The general public has been left in the dark. The news media tend to report on space science with open-mouthed awe, without asking hard questions. Instead, lavish news coverage has gone to a Dutchman with a silly scheme to send people to Mars as a reality TV show.

Within the small group of scientists and engineers studying these problems, a conversation has emerged about what happens next. On one side, the engineers ask the health experts for the standards they need to attain to keep astronauts safe and functioning. On the other side, the doctors ask the engineers for equipment to find those standards. Studying the hazards of deep space is hard without going to deep space.

John Zipay is part of that conversation as a leader of structural engineers at NASA. He feels confident that his team could design a spacecraft with artificial gravity, which would fundamentally address the risks of weightlessness. But he would first need to know how much gravity is needed.

In a career that began in the 1980s, John traveled the world to make sure the pieces of the International Space Station would fit together.

"I call it the first orbital wonder of the world," he said. "There were pieces built in Russia, and in Japan, all over Europe, and all over the United States, and we shipped them to Florida and to Russia, and we put it all together in space, and the damn thing works. It's been continuously crewed since November of 2000. We can do anything. I mean, honestly, that, to me, proved that we can build anything."

The ISS is huge. Its total length, 109 meters (358 feet), is roughly the length needed for a spacecraft rotating at a reasonable rate to create gravity equivalent to the Earth's. The crew modules are a combined 51 meters (167 feet) long and have a pressurized volume equivalent to a 747. And it was built strong. The pieces had to withstand acceleration from Earth, with forces during launch of 3 g's. Engineers also considered the impact of modules being brought together for assembly, and the impact of many dockings by the space shuttle and the spacecraft that now bring supplies and crews.

Zipay said an interplanetary spacecraft with artificial gravity could be smaller and less sturdy than the ISS, depending on the requirements the medical doctors ask for. Astronauts overcome the loss of muscle and bone with their exercises. It is possible that avoiding the eye and brain issues wouldn't require full-strength or full-time gravity. If the effects of weightlessness on the brain could be counteracted by an hour of gravity a day, astronauts might be able to get what they need in a rotating chair inside the spacecraft. If one-third-strength gravity will do the job, the spacecraft could be as small as 30 or 40 meters long. Or if weak, part-time gravity will do, maybe only the sleeping quarters have to spin.

"If you give the engineers what requirements you have for artificial gravity, based on the medical research, if it has to be one g continuously, we'll build you a spacecraft that gets you one g where the humans are," John said. But he added, "The most reasonable spacecraft you can expect, and the one that is most feasible to actually be built, is the one that requires the least artificial gravity for the least amount of time."

The idea of creating gravity by spinning all or some of a spacecraft uses a phenomenon known to any child who has played with a weight on a string. The inertia of the toy makes it want to go straight, but the string pulls it into a curve. The force needed to make the turn produces an opposite force toward the outside of the circle, the centripetal force. The strength of that outward force relates to the speed of rotation and the length of the string—increasing either produces a larger force. With a faster spin, the size of the circle can be smaller. A slower spin requires a larger circle to provide the same force.

If you are spinning human beings, biology interacts with physics. A small-radius centrifuge—for example, a one-passenger unit inside a spacecraft—could produce more gravity in an astronaut's feet than head. Doing work in that environment would be difficult, as an object, or even your arm, would constantly vary in weight as it moved. Dizziness can also be a problem. In the 1960s, an experiment spun human subjects for twelve days at a stretch. Once people are used to spinning, most can hold down their lunch, but in those experiments it took a day or two for participants to adjust, both after the spin started and after it stopped.

A solution that continuously spins the whole spacecraft gets around the problem of starting and stopping. Zipay imagines a barbell-shaped spacecraft that rotates end over end. Passengers at one end would experience 1 g. With equal gravity at either end, the craft would need to be 120 meters (394 feet) long. If it rotated around one end, with artificial gravity only at the other end, half the length would be needed. In either case, the length is dictated by the speed of the spin—four rotations per minute—suggested by the 1960s research as the fastest rate human beings can handle.

But that research is old and happened on the ground, where the centripetal force is added to the force of gravity, which never goes away. There are good reasons to wonder exactly how fast people can spin in space, where artificial gravity would be the only force they would experience.

In the current generation of space scientists, artificial gravity has received serious attention only since Mike Barratt and his colleagues stumbled on the problem of optical swelling and intracranial pressure on the ISS a few years ago. Interest increased in 2014, when Mike called together a workshop of one hundred top researchers from around the world to study the challenges of artificial gravity, or AG. The report from that meeting highlighted the gaps in our knowledge.

"Even though centrifugation forces have been known to induce AG in animals and humans for more than 100 years, relatively little is still known about the physiological effects and in particular the effects of longer duration centrifugations," the report said. "Actually, we know more about the effects of weightlessness in space than about the effects of centrifugation just beyond a few hours' duration . . . We do not know, for example, whether a Martian surface G-force of 0.38 is at all protective, and what G-threshold is needed for maintaining bodily functions during long-duration weightlessness."

Coming out of the workshop, NASA now plans a seven-year program to find the requirements for artificial gravity, aiming for answers by 2022. On its current space exploration timeline, NASA will not need the information any sooner.

Peter Norsk, a doctor leading human countermeasures at NASA, is the other half of the conversation with engineers like John Zipay. It's his job to find the information the engineers need about human requirements. Ideally, Zipay could give Norsk a centrifuge in orbit to spin people and see how they react. In fact, that's the only way to get a truly definitive answer. But such a project would be very expensive, so the current plans call for a combination of experiments with humans on Earth combined with work on rodents aboard the ISS.

Inside NASA, officials have also discussed a mission to one of the

Lagrange points, spots in the solar system where the balance of gravity from the Earth, Sun, and Moon create an area where small objects can stably orbit. Astronauts going to a space station there could learn about deep-space travel, including artificial gravity, without committing to a much harder mission to Mars. Whether those discussions lead to a new NASA direction may depend on presidential politics.

We already have a space station in orbit, but plans for a human-sized centrifuge on the ISS never worked out. In space, powering up a big spinning chair or room would send the space station spinning in the opposite direction. Controllers could potentially counteract the spin with gyroscopic wheels—spacecraft have them to control orientation and direction—but the vibration issues and opposing forces would make that impractical on the ISS.

A centrifuge large enough for primates would have been installed on the ISS but for the *Columbia* disaster in 2003, which greatly reduced the capacity to get material to the station. In 2016, astronauts will try a centrifuge for rodents, using Japanese equipment. But that will tell more about muscles and bones; rodents' eyes and brains are quite different from human beings'.

On the ground, Norsk said, the studies will use long-arm centrifuges, giving data for gravity greater that 1 g, and very short low-g experiments in a parabola-flying aircraft (the so-called vomit comet mentioned in chapter 1). But nothing in the program will give direct evidence on the long-term health of human beings in gravity levels somewhere between that of Earth and weightlessness. Nor will we know for sure what it feels like for a person to spin in space and what practical limitations go with that experience.

While the health data are lacking, the engineering solutions for artificial gravity are pretty clear.

Building an end-over-end spinning barbell in deep space is an interesting challenge, not a scary roadblock. John Zipay said it would need a power source other than solar panels, as they are too fragile for spinning. It would need to be lightweight, to reduce the energy cost to launch and accelerate toward the destination. And it would need to be strong, to withstand the forces of the artificial gravity itself.

John would reduce weight by sending collapsible pieces into

The Cassini spacecraft undergoes testing at JPL prior to launch to Saturn orbit. Note the people in bunny suits, for scale. (NASA/JPL)

Titan, with its bluish atmospheric haze layer of hydrocarbons, dwarfs the smaller moon Tethys in this Cassini image. (NASA/JPL-CALTECH/SSI)

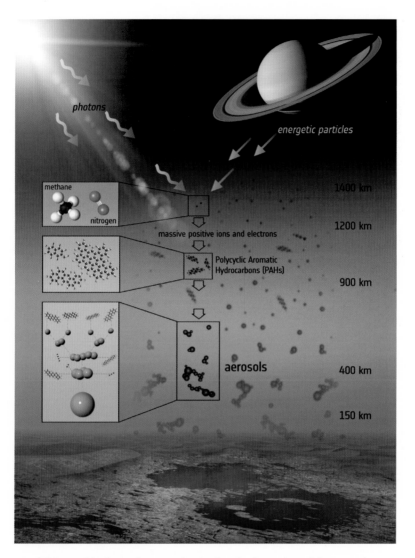

This graphic shows the steps that lead to the formation of the aerosols that make up the haze on Titan. Photons from the Sun and energetic particles from Saturn's magnetosphere interact with the upper layers of Titan's atmosphere, breaking up nitrogen and methane molecules there. This results in the formation of massive positive ions and electrons, which trigger a chain of chemical reactions that produce a variety of hydrocarbons. Many of these hydrocarbons have been detected in Titan's atmosphere, including polycyclic aromatic hydrocarbons (PAHs), which are large carbon-based molecules that form from the aggregation of smaller hydrocarbons. PAHs can coagulate into even larger clumps, which sink lower into the atmosphere. (ESA/ ATG MEDIALAB)

Methane clouds above Titan's equator in October 2010, revealing seasonal shifting of weather systems to the lower latitudes following equinox in August 2009. (NASA/JPL-CALTECH/SSI)

Artist's rendition of a methane rainstorm on Titan. (NASA/JPL/MICHAEL CARROLL)

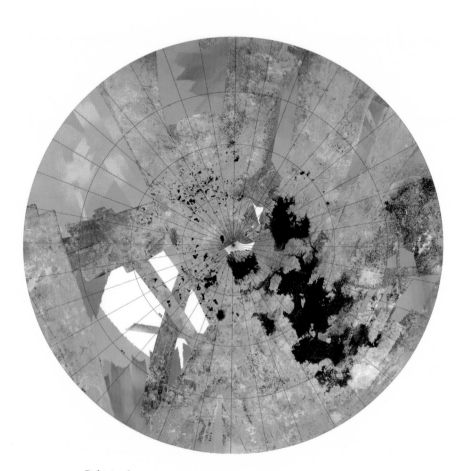

Colorized projection of a Cassini radar map of Titan's northern hemisphere, which is dominated by liquid methane-ethane lakes and seas, including the largest of the seas, Kraken Mare. (NASA/JPL-CALTECH/ASI/USGS)

Ligeia Mare, Titan's second-largest lake. This is a false-color mosaic of synthetic aperture radar images. (NASA/JPL-CALTECH/ASI/CORNELL)

Cassini's visual and infrared mapping spectrometer (VIMS) instrument acquired this mosaic image of Titan in November 2015. Some infrared wavelengths are able to penetrate Titan's thick, hazy atmosphere, allowing us a peek at surface features. (NASA/JPL/UNIVERSITY OF ARIZONA/UNIVERSITY OF IDAHO)

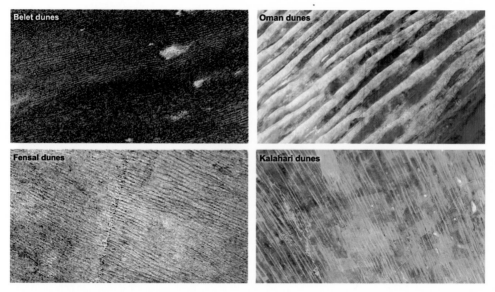

Dunes in the Belet and Fensal regions of Titan compared with dunes on Earth: the Oman dunes in Yemen and Saudi Arabia; and the Kalahari dunes in South Africa and Namibia. (NASA/JPL-CALTECH; NASA/GSFC/METI/ERSDAC/JAROS; U.S./JAPAN ASTER SCIENCE TEAM)

The Super Ball Bot tensegrity robot, in tests at NASA Ames Research Center. (COURTESY VYTAS SUNSPIRAL/NASA AMES/JONATHAN BRUCE)

Titan and Saturn as viewed by the Cassini spacecraft. This was taken in May 2012—northern spring in the Saturn system—so the shadows of the rings are seen projected onto the southern hemisphere of the planet. (NASA/JPL-CALTECH/SSI)

orbit for assembly. There is no need for all of an interplanetary craft to be built with strong aluminum segments like the ISS. The biggest part of an artificial-gravity-equipped craft would be the arm connecting the two ends of the barbell. Zipay would design an arm that could fold up into a strong, compact package for launch out of Earth's gravity, expanding in weightlessness to become a leggy truss assembly just strong enough for its purpose of positioning the two ends of the barbell. The crew compartment could be inflatable. The propulsion system could live on the opposite end of the arm from the crew quarters, as a counterweight.

Part of the radiation problem is manageable, too. The craft would have to protect passengers from two kinds of radiation. The proton radiation of solar storms is strong enough to kill, but plastic shielding can reduce its impact, and for strong solar storms, astronauts could shelter in a protected enclosure. A storm shelter surrounded by supplies of food, fuel, and water wouldn't have to be large for a radiation emergency lasting hours. This low-tech option is inexpensive and effective.

The problem with galactic cosmic ray radiation is much harder. As we discussed in chapter 5, the high-energy particles coming from exploding stars across the galaxy hit so hard they cannot be stopped by any practical form of physical shielding. Strong evidence suggests that exposure for a couple of years could damage the human brain enough to endanger a mission, as well as taking many years off an astronaut's life expectancy.

The practical potential solutions for staying healthy in the presence of GCRs are speculative.

With a strong understanding of the damage caused by radiation and the related genetics, doctors might find a way to disqualify susceptible individuals. But crew selection is tricky: screening out candidates for one risk may produce a weaker crew with respect to another risk. Astronauts could legitimately be selected for health, intelligence, physical fitness, leadership, education, technical skills, psychological robustness, radiation susceptibility, and low tendency for vision problems in weightlessness. To meet just a few of these standards, NASA's selection process already reduces tens of thousands of candidates to a few dozen. Each additional screening test would require compromises. There simply aren't enough perfect people in the world to get a crew that is the best by every criterion.

On the far end of medical science, maybe astronauts could be dosed with protective medication or cured of radiation exposure. But such a breakthrough would be akin to curing cancer. It might happen, or it might not.

Another option, if we can't survive GCRs, and we can't block them, is to bend them. The particles carry an electric charge, so their paths can be diverted by magnetic fields. With this knowledge arises the hope of producing in space the kind of intense magnetic fields we have created on Earth to bend the paths of subatomic particles in physicists' accelerators.

Rainer Meinke came from Germany to Waxahachie, Texas, decades ago to work on one of those physics machines, the biggest of all: the Superconducting Super Collider. The SSC would have shot particles around an 85-kilometer (54-mile) circular tunnel under the desert, creating subatomic collisions far more powerful than the biggest accelerators can do even today. But Congress, faced with the simultaneous cost of the International Space Station and the SSC, canceled the collider project in 1993, half built. An abandoned 30-kilometer (19-mile) tunnel remains near Waxahachie. The cancelation interrupted Rainer's career, like many other physicists', and he shifted to working on superconducting magnets, like those that would have squeezed particles in the collider, but for the power and medical industries. You may have encountered the results of that work in MRI machines and other high-tech medical imaging equipment.

Big magnets work with electrical current running through doughnut-shaped coils of wires. The current induces a magnetic field at the center of the doughnut. Using ordinary copper wire, magnets produce heat, wasted energy that has to be continuously replaced with more electricity. A superconductor can conduct electricity without resistance; no energy is lost and no heat produced. Theoretically, the current in the coil can continue forever, producing a stable magnetic field after being energized only once.

The fields produced by the superconducting magnets in MRI machines are often one hundred thousand times stronger than

Earth's magnetic field. But such magnets are much too heavy to put into space. With colleagues at his company, Advanced Magnet Lab, in Florida, Meinke developed a concept to make superconducting magnets much larger and much lighter. He would envelop a space-craft in balloonlike assemblies of gossamer-light flexible magnets made of superconducting film. The field within the balloons would bend charged particles away from the spacecraft. Another layer of superconducting coils on the spacecraft's hull would neutralize the field for the passengers inside, so metal objects wouldn't fly around.

The design is ingenious. The superconducting material, which resembles audiotape, is folded like an umbrella during launch. Once in space, charging its magnetic field unfurls the umbrella to the desired shape and size. Normally, big magnets must be very strong to avoid being torn apart by the forces they create. But Rainer's group found a new configuration of the field that would reduce the stress on the tape to close to zero. They built a small prototype, and it worked. In practice, using available materials, a system with a diameter of 10 meters (32 feet) could maintain a field of 1 tesla after a single charge, almost as strong as a typical MRI machine.

But after building the prototype, Rainer lost funding from the NASA Innovative Advanced Concepts Program that had supported the research (although he is still working with other funding and partners). Even if his system works as designed, and protects a space-craft from proton radiation from the Sun, the field wouldn't be large or strong enough to deflect the most powerful galactic cosmic rays. GCRs are far more energetic than the largest colliders on Earth have ever produced. The Earth's magnetic field doesn't protect us from them, either: water in our atmosphere does most of the work. (We are double shielded on Earth, with an atmosphere that protects us from GCRs and a magnetic field that protects the atmosphere from being stripped away by the Sun's proton radiation.)

To divert superenergetic particles away from a spacecraft, a much stronger or much larger magnetic field would be needed. Some experts discount the idea of magnetic shielding entirely, because of the weight, the complication, and the lack of a backup if such a sys-tem fails. Meinke points to the forces involved. A 10 or 20 tesla field would create massive bursting pressure on the coils, as if overinflating a balloon, forces too strong for any practical materials to withstand. The other option would be a much larger field produced by larger coils at a lower level of magnetism. That would theoretically work. But the spacecraft would be traveling surrounded by a magnetic umbrella the size of a dirigible.

The mental picture of such a spacecraft is not appealing. Add rotation to create artificial gravity, and you get a carnival of complications: a massive electrically charged superconducting blimp that travels through space at high velocity while surrounding a barbell-shaped spacecraft turning end over end, which also has a mechanism to neutralize the field in its moving living quarters. Somehow, the entire assembly is launched and propelled into space and decelerates and orbits at its destination.

Suppose that this all works and the spaceship travels for most of a decade across the solar system. What is happening inside? That scene is relatively easy to investigate. The pillownaut's loss of concentration and blah mood after months resting in a Galveston hospital bed is typical (covered in chapter 5). In 2010 and 2011, six men matching the selection criteria of international astronaut programs spent 520 days inside a simulated spacecraft in Russia. The duration matched a Mars mission, including a brief stop on the planet.

It didn't go well. In the words of the psychological study, "Only two of the six crewmembers (c and d) showed neither behavioral disturbances nor reports of psychological distress." The four astronauts had sleep problems, became lethargic, testy, and, in one case, depressed, all of them with symptoms serious enough to threaten a mission.

Similar issues have surfaced among people isolated in Antarctica and even on the ISS—although astronauts' medical details are confidential, it's said they also have coped with depression. Most people who spend months confined with a small group, deprived of sunlight and normal sensory input, eventually feel rotten and their work performance suffers. Researchers suggested after the Russian mission that the key is crew selection. They called for investigation of genetic factors that make some people immune to these effects. But that would add yet another row to the already crowded screening matrix, searching for people with multiple immunities who may not exist.

Perhaps more ideas will come along that simplify crew selection and spacecraft design. But there is already one simple and audacious solution that would resolve radiation, gravity, and psychology problems all at once: go faster.

With current technology, spacecraft carry enough fuel for a brief burn providing high acceleration before they essentially coast to their destinations. While coasting, passengers experience weightlessness. At a constant velocity, the time for the trip is set by that short burn at the start. But the math dramatically improves for a spacecraft that can keep accelerating, continuously adding more energy to its speed for the entire journey. Halfway there, the rocket could reverse, slowing down with the same power output until arrival.

During a long journey, only a tiny amount of continuous acceleration gets you to your destination astonishingly fast. Fast enough to bring passengers to the outer solar system before they lose health or mental well-being to weightlessness, radiation, or boredom, even in a lightly shielded spacecraft without artificial gravity.

FUTURE

Technology for continuous acceleration in space developed extraordinarily rapidly, seemingly overnight, like flight after the Wright brothers or the Internet in the 1990s. As in those cases, however, the ideas that made it work didn't spring from nowhere. For decades, creative engineers and scientists unafraid of working at the fringe of their professions had been thinking about far-out concepts and tinkering with prototypes without gaining much funding or respect. Suddenly, with the arrival of a pressing need came money and credibility. Conditions combined to rapidly grow the seeds of ideas, like sunflower sprouts, into something unexpectedly grand.

The CEO of Titan Corp., the independent corporation created for the Titan project, was immensely relieved when she received a successful demonstration of a functioning Q-drive. She had been promising for years a follow-up technology that would get people safely to Titan, where they could move into a habitat built by robots from slower ships that had already landed. Without her promised follow-up human mission, the huge expense of the robot missions, paid for by Titan Corp.'s funders in governments and large corporations, would go to waste.

The Titan mission architecture called for a group of six astronauts as pioneers. They would move into the small habitat prepared for them to direct robots in building larger, self-sustaining human habitats for more people to follow. They would also gather a sense of what it is like to live on Titan that no machine could express and would tell the world about it. But human beings could not survive a seven-year deep-space flight to Titan, like the flights the robots had taken and those flown by Earth's earlier probes to the outer solar system. The mission would be possible only with a continuous-acceleration drive to get there in less than two years.

A conventional chemical rocket can produce a lot of thrust, but it needs heavy fuel for its brief burst of power. A spacecraft that produces just a little thrust but keeps it up can go much faster, even if it starts much slower. The ability to keep operating comes from higher density fuel—uranium rather than chemicals—and, in the case of the Q-drive, from the ability to harvest propellant from space rather than bringing it along. The Q-drive would gather up quantum particles that naturally come into existence and expel them through an electrical field, producing thrust. A nuclear reactor would produce the electricity for the field. With that compact power source and no need to carry propellant, the drive could continuously push the spacecraft forward, slowly gaining great speed.

With the prototype working, the next steps came rapidly. Money poured in from nervous governments and billionaire investors looking both for an escape from Earth and ground-floor ownership of a brand-new world. Titan Corp. leased a large space dock and heavy-lift commercial launch rockets to build the spacecraft. It bought seats on commercial spaceplanes to get assembly workers into orbit. As the work progressed, full-sized Q-drive thrusters came together, with huge rings like detached butterfly wings to accelerate quantum particles. They floated outside the space dock.

The drives were tested on an unmanned craft with only company employees present. The crew module was being built separately. As power from the reactor came to the rings, a controller announced that the Q-drive was working and producing thrust. A cheer went up. But the spaceship itself seemed to have hardly moved. A weak continuous thrust makes for a gradual takeoff.

After the test, both Titan Corp.'s chief marketing officer and the company's chief technology officer came to the CEO's office to discuss a problem.

The marketing chief had been managing the optics of the launch to Titan. Not actual optics, as the CTO originally thought he meant, but the way this historic event would look to the public. World leaders would be on hand to see six brave astronauts depart the space dock on their way to a new home for humanity. Speeches and ceremonies would be required before the dramatic takeoff. But the CMO had watched the test and, to his dismay, realized that the takeoff would not be so dramatic after all.

The Q-drive was extremely efficient and could accelerate the craft with a continuous force that over months developed fantastic speeds. But in the first ten minutes, it would only move about 200 meters (650 feet). Watching the test, the CMO had wanted to give it a push to get it going.

This would not do, the CMO said. Waving his arms to paint a mental picture (his specialty as a drama major in a company full of engineers), he started to embellish the embarrassing scene of launching a spaceship that moved at the rate of a snail. World leaders couldn't bid the astronauts farewell, send them on their way, and then not have the exploratory ship actually leave. What a letdown when everyone finishes the ceremony, the band stops playing, and the color guard retires, and the explorers are still in sight. By the time world leaders boarded spaceplanes back to Earth, the astronauts would still be close enough to wave from a porthole. Titan Corp. would be a laughingstock, apparently sending explorers off on an interplanetary jalopy that left Earth like a baby crawling away from her mother.

The CTO was incensed and showed it as well as he could, given his inability to express his emotions, as his wife had frequently complained in their couples therapy. The Q-drive represented a leap in technology that would save humanity from being stuck perpetually on Earth. The craft his team was building would cut the trip to Titan from seven years down to eighteen months. The solar system was within reach.

The meeting ended with the understanding that the spaceship would have a first-gear rocket strong enough to get it out of sight of the space station before the Q-drive would take over. The CTO went back to his office fuming inside, grinding his teeth and devising revenge against the CMO that he knew he would never carry out. In fact, a first gear made sense to escape Earth's gravity.

The first-gear rocket the CTO's team designed would use the same powerful nuclear reactor that would electrify the Q-drive. Hydrogen in external tanks would shoot through the reactor stack to gain explosive heating and leave as exhaust. During that phase of acceleration, the big, delicate Q-drive rings would be folded. When the hydrogen tanks emptied, the spaceship would coast, at 0 g, while the collectors deployed. Then the Q-drive would take over, adding progressively more velocity through the trip.

The CMO was happy. But in a later meeting, the chief safety officer pointed out that the first-gear thruster would spew the world leaders with superheated hydrogen. An additional redesign included a tug that would take the craft clear of the space dock before it started its main engines.

The optics of a departure by tug seemed perfect. Not too fast, not too slow.

The CTO began designing a tug.

Getting the departure just right added a few years to the project, but Titan Corp. did fine. The increased cost came with a 33 percent markup.

PRESENT

At NASA, even some risk-accepting astronauts see the need for faster propulsion. If a trip to Mars is marginally doable with barely accept-able radiation exposure, getting there will be the end of the road. The history of the Apollo missions argues against pushing to the extreme end of our capabilities in a giant effort with no follow-up. After mak-ing a Mars landing at a stretch, we could repeat the fifty-year wait for the next step that happened when we pushed the envelope to get to the Moon. To continue exploring, each step should build to the following step, so we can keep reaching farther rather than waiting generations for new ideas.

Besides, the more we learn about Mars, the less interesting it seems as a final destination. We will want to go farther.

At the Johnson Space Center, Mark McDonald, who leads the Advanced Mission Development Group, explains the issue as a balance of constraints, like any good engineer. A journey to the outer solar system will require a fast flight, and not only because of radiation concerns. A long journey would require too much food and fuel on the spaceship. As the mass of the ship rises, the propulsion problem becomes even harder. A long-range exploratory spacecraft will need to be light and fast, with efficient engines that don't carry much fuel.

Perhaps fuel could be mined from an asteroid, if a promising rock presented itself and could be moved so it would stay in the right place to work as a gas station. Mark thinks the immense cost and complexity of finding and positioning an asteroid, mining it, and refining fuel would make sense only for a mature spacefaring fleet with lots of customers, not a single flight. More likely, fuel and supplies would be stockpiled on the way to Titan by robotic craft. Slower, larger spaceships without people on board could establish caches, followed by the manned mission going step by step, resupplying along the way.

But that plan carries the risk of all those starts and stops, each a rendezvous in deep space. Even if they go perfectly, they take time. Besides, Mark points out, the concept doesn't get at the basic cost issue of using chemical rockets for the outer solar system. Each additional ton of fuel lifted off the Earth and propelled to deep space is extremely expensive. And a deep-space mission would take hundreds of tons of fuel.

Today's chemical rockets are great at producing huge bursts of thrust for a short time, but they are inefficient and the fuel is too heavy for a long journey. McDonald draws an analogy to the Old West. If settlers crossing the plains had not been able to graze their horses along the way, each would have needed a long wagon train to preposition feed. No one could have afforded the trip.

Harold White, known as Sonny, works for McDonald, thinking about the fundamental breakthrough needed to build a spacecraft that doesn't require all that weight and support. He's a straightforward and personable guy with a humble manner incongruous with the amazing work he does. Sonny uses esoteric physics to come up with real machines and in the process has found new insights about the universe. But he discusses his work at the outer end of theory with the same casual enthusiasm he displays when showing off pictures on his phone. A photo of a 1960s nuclear fission rocket sitting in the desert, taken on a family trip, brings the confession, "I'm such a nerd."

But he's only a nerd if that's what you call someone on the NASA payroll who designs stuff originally dreamed up by *Star Trek*. As we'll see in chapter 12, Sonny is working on a warp drive. He's a superhero of nerds.

For everyday nerds, mortals like us, understanding what Sonny is talking about is challenging. A primer is in order.

Rockets, airplanes, and boats all accelerate using Newton's third law of motion. For every action, there is an equal and opposite reaction. A boat's motor turns a propeller that moves water backward; the opposite reaction to that movement of mass is the forward motion of the boat. An airplane engine uses the mass of the air forced through a jet or propeller to accelerate. A rocket, flying in space, doesn't have access to water or air to use as a propellant. The mass of its exhaust provides the opposite forward reaction. Regardless of its source of power and how fast its exhaust comes out, the forward motion of a conventional rocket is limited by Newton's third law to the mass of the propellant it carries on board. Boats and airplanes would be capable only of very short trips if they had that limitation.

A better source of energy and a more efficient engine can help. In the inner solar system, solar panels can produce electricity to power an engine. Spacecraft using electricity carry xenon as a propellant. Xenon is an inert gas that doesn't burn; it is only a propellant, not a fuel. In these engines, called ion drives, the electricity from solar panels ionizes the xenon atoms, stripping off their electrons to produce positively charged ions that can be moved by an electric field. The thrust comes from forcing the xenon ions to accelerate out the back of the rocket. In the outer solar system, where the Sun's rays are weak, an ion drive could use a nuclear fission reactor for electricity rather than solar cells. Reactors have been around for a long time, like the one on the historic rocket Sonny showed us on his phone.

An electric propulsion system successfully powered NASA's Dawn spacecraft, launched in 2007, which used solar energy to get to the asteroid belt and orbit both the asteroid Vesta and the dwarf planet Ceres. At the back of the engine, grids charged to 1,000 volts expelled xenon ions at 145,000 kilometers (90,000 miles) per hour. Using very little xenon, the system produced thrust equivalent in strength to the weight of a piece of paper on Earth, accelerating Dawn so gradually it would take four days to go from zero to sixty miles per hour. But gradually, Dawn made it to 39,000 kilometers (24,000 miles) per hour. The ion drive pushed it so fast with so little power partly because of its high efficiency, ten times that of a conventional chemical rocket.

But, like a chemical rocket, an ion thruster will eventually run out of propellant. When the xenon is used up, the engine stops working. That's why the Q-thruster that Sonny is working on is so appealing. It takes propellant from space, so it never runs out.

Understanding this idea requires a sojourn into quantum mechanics, which hardly anyone understands (and maybe no one does, really). Quantum mechanics is the strange physics for questions such as "Is this light made of particles or waves?" and "Exactly where is that electron and how fast is it going?" In both cases, physicists say the answer is indefinite and unknowable in any particular instance. The exact location and momentum of a subatomic particle is fundamentally a matter of probabilities, not determined facts. And that's the stuff we're made of.

The universe is a collection of probabilities. Tiny probabilities build into large objects that we perceive as real in our intuitive sense that things exist at a definite place, time, and momentum, but at the fundamental level they do not. Because our perspective on the world conflicts with quantum mechanics, the theory predicts many seeming impossibilities, such as the ability of matter to blink in and out of existence for no reason. But those seeming impossibilities have been proven real by experiment.

The name for matter appearing out of nothing is quantum vacuum fluctuation. In our probabilistic universe, virtual particles constantly flash into existence and then disappear, even in a total vacuum, simply because there is a small chance that they could. White's Q-thruster would use these virtual particles as a propellant, instead of xenon. The energy for the drive would still come from a nuclear reactor or solar cells, but the propellant forced out the back would be free, coming from space itself. Virtual particles from the vacuum present in the engine chamber would be accelerated by the electrical field. The Q-thruster uses what's already there (or potentially there) the way a boat motor uses water or a jet uses air.

White tested a version of the concept at Johnson Space Center in 2014 and produced a tiny thrust with a small amount of electricity. Although the thrust was so weak it could only be measured by supersensitive equipment, the system's efficiency was six to seven times greater than the ion drive.

"It's nothing you're going to use to lift yourself off the ground," White said. But if the Q-thruster could scale up, driven by megawatts of power from a nuclear reactor, its efficiency would make a big difference for long-distance space travel. "With that type of a system, it is revolutionary, in terms of what we've done. And it would make a Saturn mission a little less daunting."

The results are still being reviewed. Another lab needs to verify the phenomenon using a different measurement apparatus. The amount of thrust is minute, because virtual particles are few and far between, but that's manageable using the math of modern physics, too. The quantity of virtual particles that spring into existence depends on the energy state of the empty space they inhabit. Introducing mass into a vacuum increases the energy and the number of particles that appear. In Sonny's thruster, the number of virtual particles is fourteen orders of magnitude greater than what is found in the void of space, going from a number that is hard to distinguish from zero to a number that is still unbelievably small but seems to be large enough to make the idea work. (At best, it takes 10,000 cubic kilometers of space, a volume close to that of Lake Superior, to gather a kilogram [2.2 pounds] of the stuff.)

We've advanced a scenario to colonize Titan in this book with the intention that we don't want to call on fantasy to keep that hope alive. The Q-thruster is our best bet. Although it has barely been born as a technology, it seems to work. As it becomes better understood, physical laws that don't seem to make any sense may be used to even greater benefit.

"It's some speculative physics that, if we can pound it flat, might be very useful for something like Saturn," White said. "We're keeping all our options open."

FUTURE

When the ship was ready, the world watched in awe and hopeful expectation. The tug pulled the craft away from the space dock. It was an odd-looking craft with the Q-drive's big rings ready to unfold around a center crew module, designed to be as light as possible and just large enough for the six astronauts, their supplies, and a lander to carry them down to Titan from orbit. The craft was designed for a one-way trip, with the capability of becoming an orbiting station to support the colony, its nuclear reactor a backup supply of energy to recharge batteries on robots and other equipment.

The three men and three women from five nations were already friends. They had been training together for years. They knew that they would be spending the rest of their lives together, however long that might be. They expected to live a full life on Titan, but they were flying a brand-new spacecraft, essentially untested, to go millions of miles beyond the help of anyone on Earth. They were leaving behind their families, the places where they had grown up, the Earth's bright sunshine and warm air, and everything else familiar to them.

And they didn't know where they were going. They had seen the images and data sent back by the robots, but they didn't know what it would be like to live there, if they could survive the cold, the darkness, and the lack of atmospheric oxygen and liquid water. They would have to build new machines to sustain themselves. They would have to trust that their machines, their bodies, and their minds would work in that strange place.

They were unafraid. Moving off the Earth would be one of

humankind's greatest adventures and they were ready for almost anything. Except for having families.

They were very certain of their birth control methods. The future remained far away when human beings would be able to reproduce and populate a colony off the Earth.

PRESENT

With a fast trip to another planet or Titan, astronauts wouldn't need artificial gravity to maintain their health. But to give birth, they probably would.

Research on sex and reproduction in space has barely begun. Scientists, space enthusiasts, and pornographers alike complain of NASA's prudishness. After half a century of spacefaring, nothing has been published scientifically describing sex in space. Insiders at the Johnson Space Center say human sex has happened, which seems obvious, but no one talks about it officially and certainly it hasn't been studied. Sex is surely possible in space. But conception, gestation, and birth would be difficult and risky. We still know almost nothing about growing a baby in space. Human reproduction may not work without full Earth-strength gravity.

There's a ton to read about on this topic on the web and in books, but it all relies on a few old studies involving animals and plants, a handful of anecdotes from astronauts, and lots of speculation and padding. Following these echoes back to their sources— a challenge in this area of journalism, where writers feel free to use their imaginations—we found little basis for saying anything firm about how reduced gravity would affect reproduction or the growth of children. But it is a worry.

April Ronca studied these issues at NASA until it defunded the work after the *Columbia* disaster in 2003. She came back in 2013 to the Ames Research Center, as medical issues rose to the fore again, after having led the Women's Health Center of Excellence at Wake Forest School of Medicine. She begins with her knowledge base about Earth reproduction and development for clues about what would happen in space.

Environmental influences on developing brains and bodies can be permanent. For example, rodents raised in environments without horizontal or vertical lines have lifelong visual impairments. The much more extreme environmental differences of weak gravitational forces, or of weightlessness, could shape children in drastically altered ways.

"I would say it would be very difficult for a woman to give birth in space," Ronca said. Conception may require gravity. The placenta might not attach correctly. Rat studies showed that time in weightlessness interferes with birth contractions. And if all that works, what might happen to the fetus?

"If you look at some plant studies, you see that morphological changes occur in a microgravity environment," Ronca said. "I don't know what we would expect the organism to look like, going through those thoroughly embryonic pattern-formation stages and organogenesis. How do things get hooked up properly?"

Four American male astronauts reported a lack of interest in sex and reduced testosterone, but a busy, stressful life in orbit could cause that, and the sample size is too small to draw any conclusions. The fluid shift of weightlessness produced unwanted and even painful erections for some astronauts, so there's no need for Viagra on the ISS. The problem of copulating in weightlessness has been discussed for decades. Without gravity, it would be difficult to create the right forces for penetration and thrusting. In 2006, a minor film actress and science fiction writer named Vanna Bonta invented a suit to solve this problem, allowing astronauts to Velcro themselves together. The idea brought her a mother lode of free publicity and lives on as a popular media meme.

Sex only requires friction, a simple mechanical problem. But human conception has never happened outside Earth's low-radiation, 1 g environment. Researchers think that the radiation and weightlessness of long-range exploratory spaceflight could cause temporary infertility for men and possibly for women. It doesn't appear to be permanent. No follow-up studies have been conducted, but male and females astronauts have had healthy children after traveling in space.

Like many complex biological processes, reproduction uses gravity. Experiments at the University of Montreal showed that low gravity affects plant reproduction. Chinese scientists found low gravity and radiation damaged sperm in mice. In the 1980s and '90s, U.S. and Russian scientists studied reproduction of fish and various animals in space. The fish did OK, but rat embryos showed decreased bone mineralization and shriveled heart ventricles. Rats born in space did not behave normally. No mammal has yet been conceived and born in space.

It's unlikely humans will ever attempt to reproduce in weightlessness. Even if it proves possible, there's no reason to take the enormous risk or go to the expense of creating an operating room or nursery to perform weightless Cesarean sections, since natural labor probably wouldn't work. For example, Ronca pointed out, getting fluid out of a newborn's lungs would require technology no one has even thought about inventing. Without artificial gravity, women who get pregnant in orbit would need to return to Earth as soon as possible. On long-range missions they would need foolproof birth control and the ability to have abortions.

But to create a colony we will need children. Unfortunately, studying conception, birth, and growth in a low-gravity environment is even harder than studying reproduction in weightlessness. The lack of a centrifuge on the ISS holds back the research, as it is impossible to create fractional gravity on Earth. Ronca and her colleagues are using centrifuges to study developmental issues in animals in gravity greater than 1 g. The body's response to gravity seems to obey a dose-response pattern, so it's hoped that this curve could be extrapolated below 1 g. But there could also be thresholds we don't know about for big changes in the way biology responds to gravity.

Speculation varies as to how children would grow in reduced gravity. Experiments in the 1990s found that litters of rats failed to nurse and only survived because astronauts intervened. The experiments didn't last long enough to learn about development. Bones determine our size, and bone cells align under stress. Less gravity would mean less stress and less force to orient cells, producing bones in altered shapes, perhaps too short or too long. Certainly, children would be weaker, as much of our muscle strength on Earth also develops from resisting gravity, including in our hearts.

"We don't know if you would even get to the point of having

elongated bones," Ronca said, because the reduced gravity might interfere with essential developmental stages. "Just thinking about all the things that can go wrong, it's hard to imagine. If everything went right up to the point of early postnatal development, perhaps, things are moving in the right direction, and you see the effects of gravity changing bone. Whether they would be longer or shorter, I don't know. I think they definitely would be weaker. And I think the brain is potentially going to be shaped differently from developing under a headward fluid shift."

But human beings can compensate for big physical differences. As a graduate student, April studied children born without one of the hemispheres of the brain and adults who live normal lives with large parts of their brains missing.

Possibly, children growing in low gravity would remain healthy as long as they stayed in that weak gravitational field. Many writers have said so, although there's no scientific basis for that conclusion. But they probably could not return to the full-gravity environment on Earth. Their bones and hearts would be too weak. Today's astronauts avoid bone and muscle loss through an extreme workout regime or recover it after they return to Earth. Maybe space children could be conditioned for trips to Earth, but likely not, since their permanent growth would have been shaped by the low-gravity environment.

Conception and gestation could be assisted technologically. We don't need sex for conception even today. Pregnancy could progress under artificial gravity. It would be possible to spin a woman for nine months on a space station or in a centrifuge on the surface of another planet. If you think pregnancy is fun now, imagine it in a centrifuge!

But even if that works, making the decision would be huge—an even bigger decision than it always is to have children. A spacefarer's children likely would never be able even to visit the Earth. With children, a space colony would have to become permanent and self-sustaining.

8

THE PSYCHOLOGY
OF SPACE TRAVEL

Scaring an astronaut is difficult. Loredana Bessone, a trainer at the European Space Agency, discovered this problem when she had the idea of putting astronauts in a cave to create stress, an important part of any realistic training. While working underground to develop the idea, a psychologist in her group was left behind for forty-five minutes and, sitting alone in the dark, saw her life pass before her eyes. Loredana described the psychologist's experience later: "I was alone, there was no sound, there was no light, there was nothing around me, no smell. So I started feeling like, nothing. It is a feeling of nothingness, other than your own breath, which is terrifying."

That seemed perfect. Loredana re-created the experience for the astronauts, all alone in total darkness, helpless, with no idea how long they would be left behind.

The astronauts took naps.

"They said it was a wonderful opportunity to sleep, so why lose it?" she said. "I stopped doing the exercise, because it was useless."

But she had found a perfect place to train astronauts, in a cave under a remote, undeveloped valley on the island of Sardinia. There are lots of ways to simulate teamwork in space, and many so-called analog missions have been tried—in a Hawaiian volcano, on a Canadian island, or in a can-shaped enclosure inside a large workshop at Johnson Space Center, among others. But that's always pretending. In the cave, astronauts explore and map new passages, take scientific samples, look for new life, and probe ever deeper and farther from the surface, a hazardous mission without easy communication or immediate rescue. They sometimes squeeze through gaps in the rock so tight they have to exhale to avoid getting stuck.

Francesco Sauro, an expert caver who instructs the course, notes that no astronaut ever refuses to go through a tight gap if told someone else has done it before. They're too competitive. As a veteran astronaut told us, they would rather blow up than mess up. But Francesco has also noted that they consistently overestimate what they can do in a day, underestimate their need for rest, and after a few days in the cave have to reconsider their limits and sleep-wake cycles. They are used to using systems, timelines, lists, and formal procedures.

The professional cavers are different. They're independent. They're good at adapting and improvising. They're scientists and explorers seeking a goal, but easygoing, not focused on rules.

For deep-space exploration, astronauts may need to learn to be more like the cavers. NASA has long lived by checklists and the strategy of trying to prepare for anything that can go wrong in advance. Astronauts have never been so far from Earth that they cannot ask mission control what to do in real time. But checklists might not work for exploring another world. The time it takes for a radio signal to get back to Earth will make conversation with controllers impossible (communication will be more like e-mail or text messaging). The unknown of Mars or Titan might be a lot like the unexplored pit of a dark cave.

Astronauts recognize that, and they love the cave. Bessone's CAVES program is so successful that all the nations participating in ISS missions are rotating their cadres of astronauts through, even the Russians and Chinese. (The acronym stands for Cooperative Adventure for Valuing and Exercising human behavior and performance Skills.) Trainees are left on their own to explore and, while learning to work with other nationalities, have to choose their own leaders. Five or six astronauts go down, each initially with a guide—the first

part requires a dangerous, technical descent into the cave network—
and explore underground for six days at a time. A film crew records
their reactions to everything that happens, but the project promises
confidentiality. For adventurers, it is a challenge and an amazing
opportunity.

Astronauts learn about working as a team against real hazards
and unexpected conditions. The cave is like space because it is objectively unfit for humans. As on a spaceflight, astronauts feel isolation,
confinement, and lack of privacy. Facing authentic risk, with poor
communication, they have to carry out real field activities. Distant
from family, they're forced to live, collaborate, and share space and
resources with new coworkers. It is an alien environment where people can't live without technology.

The project is for training, not research, but the practical lessons
of operating in the cave seem applicable to planetary missions. Technology breaks in the cave. Far from help and in a hostile environment, everything needs to be as simple and robust as possible. And
astronauts usually are not field scientists. Outside of their own areas
of expertise, they know as much as interns or, if trained, lab technicians. It takes an education and career to become a scientific expert
who can pick out new and important discoveries from a landscape.
The Apollo program sent a professional scientist to the Moon only
once, the geologist astronaut Harrison Schmitt on Apollo 17, and
that was the mission that also gathered the most meaningful scientific data.

So far the astronauts have mapped about 5 kilometers (3 miles) of
the cave. It could go on for tens of kilometers through the limestone
karst. In 2012, they found a new species of woodlouse in the cave,
and they take microbiological samples for use in serious research.

Handling information will be a key challenge for space explorers.
They can't possibly know everything they need to know. In a small
crew, it might make sense for each astronaut to be able to treat the
others for physical or mental illnesses. But they can't all be doctors
and psychologists. No one can know everything known by medical
professionals, engineers, geologists, technologists, pilots, and spacecraft repair people. And that leaves aside the enormous information
load required just to carry on daily life in space.

Travel for a long time in space is difficult. Whatever the original justification for building the International Space Station, it has demonstrated that fact, while teaching countless lessons on how to manage a long space voyage. It is also an analog mission.

Some of the most valuable lessons from the ISS have been unexpected. For example, no one realized how difficult it would be to store and find stuff on board. Anyone who has traveled on a boat knows how important it is to put everything away in a certain place. The ISS is like a boat that never returns to port, and the astronauts spend much of their time putting away supplies that come up from Earth, packing garbage in the capsule to return, and keeping track of items stored in countless cubbyholes in the various modules.

Designers of the ISS didn't come close to anticipating how difficult the storage problem would be. At first, the accumulation of stuff was overwhelming and the station a perpetual mess. Now every item is logged into a database as part of a detailed process of keeping track of everything. Even at that, at least one crew member has received a cell phone call after getting home, asking for help finding a lost item on the station. The cost is in time, or overhead, as NASA calls it. After the hours spent eating, sleeping, washing, system maintenance, the two hours a day of exercise, and all the time it takes to put away and look for stuff, astronauts average only thirteen hours a week of actual work on their mission.

One way to address the problem of managing objects is to take less stuff along. A 3-D printer will help fabricate parts in space, requiring fewer spares. The ISS astronauts' exercise equipment uses up a lot of spare parts. The machines are unique and complex, with many moving parts, allowing long workouts under stresses that simulate gravity while at the same time isolating the vibrations of exercise from the low-g experiments on the station. Pieces wear out. The printer could create spares when they are needed, so the ISS would only need data files, not actual objects. That strategy would be even more valuable on a mission to another planet, with resupply impossible and storage space even more limited.

Marc Reagan is studying a similar concept for giving astronauts the knowledge they need on a novel, long-range mission. He works in a NASA office that runs analog missions and he is also a part-time

CAPCOM, or capsule communicator, in the ISS control room at Johnson Space Center in Houston. The CAPCOM is the primary person who talks to the astronauts. Marc wants to reduce the importance of that role—and the dependence of astronauts on mission control—by putting more information on board.

The control room at Johnson Space Center looks just as any science fiction fan would expect, with rows of consoles facing a big screen that shows the path of the ISS and the satellites communicating with it on a large wall. The main function of the room, and similar rooms in Russia and Japan, is to communicate expert knowledge. Staff members at each desk monitor an aspect of the flight and provide direction through the CAPCOM to the astronauts, giving them the benefit of a broad range of expert advice.

NASA has learned over decades of spaceflight that conflict naturally develops between the crew and the people on the ground, as the astronauts receive controllers' many demands and perspectives and feel frustrated with a lack of understanding on the ground of exactly what they are experiencing. The CAPCOM is the translator and advocate who smooths out the relationship and regulates the flow of information.

On the ISS, astronauts can surf the web and make private phone calls to anyone on Earth. Flying is much more like a job than an exploratory mission would be. Houston wakes them up, works through the day with them, and puts them to bed. Every minute is scheduled on a timeline. Procedures are documented with written step-by-step instructions. The NASA tradition of leaving nothing to chance creates a culture that, at its best, is about precision and discipline, with a team of well-trained people carrying out carefully laid plans with perfect fidelity and without need to improvise.

As a spacecraft leaves the Earth on its way to Mars or beyond, that culture will have to adapt. Exploration is about the unexpected. It cannot be scripted. Astronauts will have to be more like cavers. They will need ways of applying knowledge to new situations without having instant communication with a team of experts.

Reagan's analog for studying these issues is underwater off the coast of Florida. Astronauts dive to a laboratory operated by Florida International University, 11 kilometers (7 miles) off Key Largo and

19 meters (62 feet) below the surface, and stay there for about two weeks. The underwater habitat is much like a spacecraft and the dives are like spacewalks, with the ability to adjust buoyancy to simulate weightlessness or the fractional gravity of Mars. Like the cave analog, Marc's program, called NEEMO (NASA Extreme Environment Mission Operations), involves real risk, isolation, and challenges. Astronauts' bodies even mimic some spaceflight changes, like the phenomenon of viruses activating under the stress of the mission.

To take NEEMO's analogy beyond low Earth orbit, Marc introduces a communication delay between the lab and controllers on the surface. Deep-space exploratory missions will go beyond the reach of instant communications. Light-speed radio signals take four minutes each way traveling to Mars, when it is at its closest, and nearly ninety minutes to Titan. The model used for controlling the Mars rovers—of sending a day of detailed instructions all at once—will not work for human beings. Marc doubts the current form of NASA checklists and timelines will work, either, as inevitable misunderstandings and unexpected situations would require back-and-forth clarifying messages that would be frustratingly time-consuming and inefficient.

"That's the challenge we will have as we go from this mission-control-centric model we've had for fifty-five years, to a more crew-centric model where they're out there on their own and have to solve things as they come up," he said.

Somehow, the astronauts will need more knowledge than they could possibly receive in training. Marc believes video offers the solution, similar to the way he uses YouTube videos to learn music on the piano or to fix a broken taillight on his car. He has used a video for that.

"I can't get the darn thing off without breaking it," he said. "But fortunately, someone has posted a YouTube video. And I don't have a ten-page instruction manual, but that fifteen-second video is all I need."

Loading a spacecraft with thousands of videos covering myriad tasks and situations would give the crew a bank of information they could quickly absorb and give them autonomy from the experts at mission control. Marc sees home improvement television as an

inspiration. Those shows don't depict the entire process of installing a faucet, just the few seconds that are tricky, covering three-
dimensional quirks that would take forever to explain in writing. A
video on repairing a complex piece of a spacecraft could be prere-
corded by experts on Earth and edited with narration to give only the
highlights an astronaut might need to do that task on a flight.

The underwater missions and the ISS provide a setting to test
these ideas, as Marc and his colleagues learn to make videos that
really work. Astronauts take equipment down to NEEMO and try it
out before it flies. For example, a heart rate monitor worn by astronauts on the ISS kept dropping large amounts of data; NEEMO
tested a newer, Bluetooth version to gain confidence it would work
in the similarly sized metal enclosure before sending it up. The
underwater astronauts also tested a drill that could take core samples from the surface of an asteroid, working with neutral buoyancy
below the ocean that replicated the weightlessness of space. With
the communication time delay, these tasks modeled deep-space challenges. Astronauts had to think for themselves and rely on pre-
recorded guidance.

The work is interesting and rewarding. Reagan was a crewmem-
ber on a nine-day mission himself. "That was about the most fun
thing I've ever done. It was a blast. It was over in a blink of an eye. I
was really sad to leave."

Astronaut Barratt had similar feelings about being in the Italian
cave. "You can't imagine how beautiful it is down there," he said.
"When you're down there, and you're feeling that cool cave air, and
you're hearing the echo of a drop that could be kilometers away,
and you're seeing this incredibly beautiful place, and you're realizing
how few people have been there. It's pretty awe inspiring."

Similar feelings might carry space explorers through missions
of a year or two without psychological breakdowns. Being among
the first human beings to fly to another planet would surely provide
enough excitement to keep focus and a positive outlook, even for a
year in a small capsule. Returning toward home would carry another
kind of excitement.

By then, astronauts would be fully adapted to their new environment, and getting used to life on Earth again would be a new challenge. Francesco Sauro said it takes only two or three days in the cave

before the mind and body adjust to using energy and operating efficiently in the dark, cramped environment—similar to the adaptation
of astronauts to weightlessness. And the senses quickly forget about
the odors of the outside world and become hypersensitive, something astronauts and cavers both recognize when they return to the
normal world.

"You smell earth," Loredana Bessone said. "You smell leaves. You
look at a tree, and you smell the leaves of that tree. It is amazing. I
never felt that before being in a cave for six days. You look at the
earth and you say, 'That smells. That has an odor that I never felt
before.'"

That vividness lasts only fifteen minutes before the normally
duller, mixed palette of the senses returns.

Researchers in Antarctica report similar sensory deprivation.
Some pack curry for the trip and cook spicy meals to compensate.
Workers who have overwintered there experience a strange divorce
from the smells, tastes, and social world of the rest of humanity.

For the life of explorers living on Titan, and the colonists following them, Antarctica is the best analog.

FUTURE

The six Titan explorers would check in with Earth each morning,
part of the daily routine designed to keep them linked to normal
sleep-wake cycles during their journey. Imperceptibly, the conversations slowed down, with a longer and longer wait for an answer to
each transmission. The point at which the lag became unbearably
frustrating differed for each astronaut. No one wanted to talk when
the wait for an answer passed twenty seconds, less than 4 million
miles into the trip, and half of a percent of the way to Titan. As the
Q-drive accelerated the spacecraft, the length of the lag grew more
quickly. The astronauts woke to e-mail messages from Earth, which
they answered at their convenience, knowing they were well beyond
the reach of disapproval from mission control.

The Q-drive didn't accelerate rapidly enough to produce strong
artificial gravity, but people and things did tend to slowly sink toward
the back of the ship. An astronaut could easily stay in place by linking

an arm through a railing and with a leap could fly from one end of the crew commons to the other. But an astronaut who floated free, forgetting to resist the spacecraft's gentle force of acceleration, would slowly sink to the effective floor of the spacecraft, coming to rest among the tools, wrappers, clothing, and anything else that had been left to floating. The phenomenon made cleanup easier. Spills all went to the same place.

A siren screamed through the spacecraft a month into the voyage. Earth had detected a solar flare that would bathe the travelers in dangerous proton radiation. The astronauts gathered reading matter and snacks, watching the clock for the flare's arrival, then flew and crawled to the spacecraft's storage bay, where a shelter had been carved out among totes of food and tanks of water. With the trip near its start, most supplies had not been used up, leaving a small refuge into which the six could fit only with uncomfortable closeness, lacking enough room even to hold up their e-readers.

During the final phases of training these six had learned they would be traveling together to Titan. Two heterosexual couples formed within a week. (Everyone had been dosed with long-lasting birth control medication to prevent pregnancy for at least a few years.) The third woman made it clear she was not interested. The third male astronaut laughed at his predicament. He was an easygoing guy, a gifted engineer who could fix anything, and not strongly driven, sexually or competitively. He joked, "I'll clean up when they break up." He wasn't really worried about the social connections of the others or particularly aware of what they thought of him. His mind focused on tinkering and improving the spacecraft.

There were plenty of maintenance jobs. The nuclear reactor needed monitoring. Navigation required small daily adjustments. The tinkerer also looked for ways to adapt the spacecraft, getting its computer to talk with contractions—saying "it's" rather than "it is"— and piecing together a still to make alcohol. No one mentioned that project to mission control. The other astronauts enjoyed traveling with the tinkerer, because he was always cheerful and had ideas for how to improve their lives, and, if in the right mood, they could even enjoy his dumb jokes and puns, with a groan.

The recreational work stopped when the flies hit. Everyone on board had been sick for the first few weeks with illnesses ranging

from sniffles to flu. Now they found themselves being buzzed by little black flies. The spacecraft was small, but they searched high and
low for the source, until one of the astronauts, an engineer, confessed
that he had brought a potted azalea along and had been nursing it in
his gear locker.

An unsterilized organism—a plant and soil, with flies—violated
Titan Corp. procedures. Initially the astronauts planned to freeze
and eject it. But when they looked at the plant with its small green
leaves and pink flowers, they were transfixed. It was indescribably
beautiful, rich with fragrance, and fascinatingly shaped—twisted and
unpredictable, unlike the entirely mechanical world that surrounded
them.

The tinkerer began studying how to kill flies on a houseplant.
The spacecraft carried none of the chemicals necessary. He began
a chemistry project to make poison, without support from Earth.
Mission control ordered destruction of the plant and said untested
chemical mixtures were too dangerous.

The fly extermination project seemed successful, but the flies
always came back. The astronauts would talk about seeing flies at
their weekly bridge game—and trade humorous accusations about
who was responsible for harboring the pests on board. They had
learned it was better to work by themselves most of the time rather
than stick together and try to talk after they had run out of things to
say. The Saturday night bridge game, with a few sips of grain alcohol
from the still, was the best time to swap conversational nuggets they
had developed during the week and share the personal issues they
were facing. After the game, everyone felt a bit more normal. They
looked forward to the next week's game.

Only the mission commander declined to participate in the
game or imbibe the alcohol. He was more intense and driven than
the others and felt responsible for the mission, upon which, he often
reminded himself, the fate of humankind might depend. The female
astronaut who had paired up with him in training, a robot programming and control expert, gently broke up with him. He was never fun
and she felt too hemmed in by his constant presence. The weight of
isolation on the spacecraft was bad enough without trying to cheer
him up.

The mood dragged for everyone as the mission progressed, but the astronauts made an effort to stay positive in their interactions with one another and to look forward to Saturdays. The commander, however, kept more and more to himself, getting up and going to bed much earlier than the others and holding back his irritation at their casual attitudes only with obvious effort. He maintained a strict regimen of exercise, according to the schedule necessary to remain fit for return to Earth, working out ninety minutes a day. The others cut back their workouts to a few times a week.

The astronauts groused about the orders that came in numerous daily e-mails from Earth. They became less and less concerned about following official procedures. But the commander took the side of ground control. He scolded anyone who deviated from the rules or skipped exercise periods. No one confronted him directly, but they stayed relaxed and did what they wanted to do, sharing their annoyance over the bridge games.

The commander oversaw the turning operation, calling everyone to duty stations, strapped in and ready at controls. Halfway to Titan, the spacecraft would gradually turn around so that, for the second half of the trip, the Q-drive would be slowing rather than increasing its enormous speed. The computer did the work and the astronauts inside the spacecraft barely felt the maneuver. The tinkerer used the time for a nap.

The bridge games ended as Titan grew nearer. The cards had worn out. The realization that they could not be replaced hit people hard. For the explorers, Titan would be survivable only with resupply from Earth. Most things they brought from home were irreplaceable. A self-sustaining colony was years away. The journey had seemed to be the largest challenge, but now the astronauts realized that landing would only be the beginning. The hardest part was ahead. Totally alone, they would be solely responsible for the machines that would keep them alive.

The spacecraft had been loaded with thousands of videos to cover all kinds of repair issues, medical problems, and even difficult relationships. An artificial intelligence application found and served up the videos, using context to guess which one was needed. The computer also contained a series of artificial personalities that could be embodied on a screen to carry on a conversation with the astronauts, adapting to their concerns to become good companions. The idea was to provide a variety of interactions so life with the same six people wouldn't get stale. But the personalities and motivations of the computer's faux companions were too easy to guess and their answers became predictable too quickly. Besides, they had nothing on the line; they couldn't really share the astronauts' lives. The crew played with them, making strange comments to cause their computer friends to adapt in weird ways.

The spacecraft slipped smoothly into Titan's orbit. As they approached the area where their habitat had been prepared by the robots, the astronauts boarded the landing module and launched from the mother ship. Descending through the thick, dense atmosphere, they set down softly. A cloud of reddish hydrocarbon dust rose outside the windows in the dusky brown light.

The commander entered the airlock in his heated suit and respirator, verifying that cameras were running before he stepped out and said, "A small step for a man, the next step for humankind." He had worked on the phrase for months but still wasn't quite sure it had the right ring to it. When the video arrived on Earth ninety minutes later it was replayed, dissected, and discussed worldwide, but the astronauts never heard any of that. They felt alone.

The commander's clear mask immediately frosted up on the inside, as the moisture from his body hit the supercold plastic. He couldn't see anything. He tried warming the mask with his gloved hand. He even lifted it off his face to scrape the ice—the atmosphere outside smelled nasty and the cold bit his skin, but it did him no harm. Heaters were supposed to keep the visors clear, but real-life conditions differed enough from the tests that the system didn't work. The icing would not stop.

The commander re-entered the spacecraft. After a discussion, the tinkerer came up with the idea of wearing a second mask on top of the first to insulate the plastic from the cold. The commander sent a message to Earth requesting approval for the idea, but the modifications were complete before an answer came back. The astronauts emerged from the ship and looked around, feeling the awe of being on a new world. They were already on their way to the habitat when a message from Earth arrived asking for clarification of the commander's question about the mask.

The commander led the way toward the habitat, which they could see less than a kilometer away, its inflated skin glowing like a lighted yellow jelly bean. The robots had turned on the lights anticipating their arrival. The distance seemed to take forever to cross. Walking on Titan, with its weak gravity and thick, cold air, was like walking at the bottom of a swimming pool. Everyone seemed to move in slow motion. The tinkerer tried leaping but floated down so slowly that he made no more progress than the others. As the long walk continued, he thought about attaching a propeller to each astronaut to help them scoot along or fly. Otherwise, walking so slowly would drive them crazy.

They arrived at the habitat tired and ready to take off their thick suits and rest. Entering through double doors they flopped down in the small common room, a space that was strangely familiar from training exercises on Earth but that now seemed amazingly warm and bright after their time under the dim, diffuse light of the frigid Titan surface. Inside, the air contained oxygen instead of methane, but mostly it was nitrogen, as it was outside and on Earth. However, the warmth inside thinned the density of the air, making it less resistant to their movements, so they could walk faster. But on removing their masks, the astronauts realized the oxygen system wasn't working properly. The air reeked of ammonia and stung their eyes. They put their respirator masks back on.

The tinkerer said, "Welcome home everyone."

PRESENT

Colonists off the Earth will get depressed, their immune systems will suffer, and they will have trouble eating and sleeping. Lawrence Palinkas, known as Larry, a professor at USC, studied workers who stayed the winter in Antarctica for insights about astronauts going to Mars. Back when he did his most intensive work, overwinter stays in Antarctica were a lot like visits to another planet, with life in a self-contained station, hostile conditions outside, no natural light, and little communication with the outside world (although indoors the stations are perfectly comfortable and the food is ample). Antarctica is far less isolated now, with Internet and phone links and flights possible through more of the year. But overwinter syndrome is still common.

"Most individuals experience mild to moderate psychophysiological disturbances after several months of winter confinement with symptoms such as insomnia, irritability and aggression, anxiety, depression, impaired cognition, reduced motivation, gastrointestinal disorders, and musculoskeletal aches and pains," Palinkas wrote, summarizing his research and others'. "These symptoms appear to increase over time during the winter, peaking at mid-winter."

Legendary psychological crack-ups happened in the 1950s, when year-round occupation of Antarctica began, including a near mutiny against a hated leader and one resident who became psychotic and had to be locked up for the duration of the winter in a room padded with mattresses. In the early 1960s, psychological screening began eliminating people with pre-existing mental health problems, and now incidents are rare and, when they happen, are usually fueled by alcohol. A resident might get violent over a bad romance, sink into drunken lethargy, or despair about an illness or death back home.

But mostly, people just mope and exhibit the polar stare, the disengaged gaze of someone who has turned off his feelings. The social tedium and lack of sensory stimulation drive many normal people into a funk. IQs decline 5 to 10 points. Space doctor Christian Otto, who spent two winters in Antarctica working as a physician, said weeks of encountering unsmiling faces can make it startling to Skype a colleague at home who shows normal emotional responses.

Palinkas said it is relatively easy to screen out people likely to run into serious problems (with the exception of alcoholics, who often know how to hide their addiction) but much more difficult to screen *in* people who will do well. He essentially gave up trying to pick people immune to the malaise. Even if he could do it, he said, NASA is already screening potential astronauts for too many other parameters to also try to find the rare person who doesn't get depressed in a dark, lonely place.

Getting through the seven-month winter in Antarctica calls for a personality opposite to that of a human resource director's ideal employee, judging by the list of qualities Palinkas came up with. Extroverts, achievers, people who are highly motivated and value order, who enjoy affection from others and like their friends to be efficient—these are the people who are more likely to get into trouble. People who accept situations as they are and don't mind good-enough solutions, who are introverted and don't need social contact for support but get along well with others, and who try to get things done but don't worry much about when or how—these are the flexible survivors. Space colonies need easygoing improvisers, not standard-issue class presidents or Eagle Scout astronauts.

Dale Pomraning fit the bill perfectly. He was overwintering at the McMurdo Station in Antarctica with about 125 others in 1988 and 1989, when Palinkas was doing his work. A quarter century later, they remembered each other. People who overwinter tend to remember their co-inhabitants, even if they don't keep in touch. Dale recognized a McMurdo coworker he hadn't seen for two decades by the way he walked in the grocery store. He remembered Larry for telling him that colleagues had often listed him as a person they would like to overwinter with again. Larry remembered Dale for the same reason, and for his attitude. Dale focused on his job and didn't engage in the station's epidemic of gossip, which kept most residents hiding in their rooms.

Dale lives in Fairbanks, Alaska, now, a place as cold in winter as Antarctica, and builds scientific equipment as a machinist at the University of Alaska's Geophysical Institute. He's known for being able to build anything and he's proud of it. In 1989, when the wintertime loss of a satellite tracking dish threatened the station's work, he fixed it with a piece of a metal bed frame and bolts scavenged from the tracked snow vehicle that he rode to the antenna's site. He admired the Seabees, the navy construction guys who razzed one another constantly, and the plumbers, who would work on frozen sewage pipes in extreme cold without complaining. He had contempt for the lightweights who griped about the cold but wouldn't bundle up.

Dale loves to tell stories and his winter at McMurdo gave him plenty. When the last plane left in February, he recalled, the dozen women at the station rapidly found boyfriends for the rest of the year—Larry Palinkas calls that "adaptive," because romantic relationships help people make it through isolation in extreme environments. And Dale recalled how others would push him to drink at parties, although he has never been a drinker—Larry said alcohol has always been part of life in Antarctica, and most people do OK with it (as Russia's cosmonauts do on the ISS). Dale's suitemate fell apart due to binge drinking and locked him out of the bathroom for days until managers intervened. In general, Dale enjoyed his work, engaged with people who could take a joke, and didn't get caught up in conflicts or drama. The winter went fast.

"Some people really didn't like the darkness, so they would just sleep all the time," he said. "Myself, I'm a night owl. So it didn't bother me, I would just stay up a lot more. It was kind of funny. I didn't really miss the sunlight. But what happened was that time shifted. A week seemed like a day, and then a month only seemed like a week. There was this whole change of time, it just seemed to pass so fast."

Most people don't feel that way. Their mood sinks. Strict routine can make it worse. Work is the best time killer, but for goal-oriented people it can feel frustrating, as things break and cannot be replaced, tools get lost in the station, and supervisors give unreasonable orders from far away without understanding the difficulties.

Even without getting depressed, talking to the same people every day can be difficult. Andy Mahoney, who enjoyed studying sea ice over a winter at the small New Zealand station at Scott Base, said he came to appreciate the company of people who didn't say anything and didn't expect him to talk.

"You start having a conversation—and this is a common phenomenon that people observed—that conversations would just peter out in the middle, and no one was quite sure who stopped talking," Andy said. "It was almost like you had had all the conversation you could have. All the small talk was eaten up. And having proper, thoughtful conversations required engagement and energy to do so, and sometimes you didn't have that."

Weekly visits by a group from the American station for a drink at the Kiwis' self-serve bar helped everyone stay alert. Talking to new people reawoke social impulses for workers who had spent too much time together. Antarctica bases have a tradition of a midwinter party, with special food and dress-up clothing, that lightens the mood at the time when it tends to hit bottom.

For Titan colonists, there probably won't be a mid-journey break. They may be gone for good. Or, if they return, it could be years later, when their lives at home have been erased. If you have been away from home for a long time, you may know the feeling. When you return, you get to see how easily life has gone on without you.

Dale Pomraning spent a year in Antarctica, then rode a mountain bike he had built through New Zealand and wandered around Australia, getting home to his parents' farm in Minnesota two years after he had left. While he was in Antarctica, his father had suffered a heart attack and had surgery. At the time, all Dale could do was tell himself that he couldn't do anything about it and that worrying wouldn't help. But when he got home, the changes sank in.

"My parents had sold a lot of the land, and there was more strip malls and stoplights and more traffic," he said. "I got really just kind of bummed out, and I did not know what I wanted to do anymore. I did not want to be stuck in this boring, mindless machine shop, because I'd been outside in the world, doing some interesting stuff."

A friend Dale had made in Antarctica called from Alaska and that was how he ended up living in Fairbanks. From there he got hired to go back to Antarctica, drilling ice cores at the South Pole. He went nine more times. If dark icy places were a good analog for a colony, he had found his home there.

FUTURE

The robots had done their best to make the astronauts' arrival welcoming, but, being machines, they missed a few things. First, the colonists had to make the place habitable, then make it comfortable. After they had taken care of themselves, work could begin building the colony. Everyone was busy with novel and important tasks for the first several weeks. Being part of a skilled team working hard and selflessly was effective medicine to stave off homesickness and depression.

The tinkerer made his first goal to modify the system that turned Titan's ice into breathable oxygen so it wouldn't produce the harsh odor of ammonia. The commander wanted him on other work, and

Earth agreed. Mission control insisted the ammonia level was safe and the colonists should ignore it. But the tinkerer said that was ridiculous and kept at the job with the support of the other colonists. An emotional fracture deepened, with the commander and Earth on one side and the other five colonists on the other.

The colony's mechanical center occupied a building a few hundred meters from the main habitat. It was fed with resources three ways. The Keystone Pipeline brought liquid methane from Kraken Mare. An intake vent pulled in nitrogen from the atmosphere. And a conveyor delivered ice rubble from a site where a mining robot diligently cut and crushed chunks of water ice from the ground. Coming out of the mechanical center and leading to the habitat were electrical power cables, a duct for breathable air, and a pipe for drinking water. A smokestack disposed of waste heat and gases from power generation.

The ice conveyor entered the building at a heated chamber that melted the ice and fed water to two systems. One system purified the water for drinking and household use. The other performed electrolysis, exposing the water to an electrical field and splitting oxygen and hydrogen out of it. Most of the oxygen supplied a draft to burn methane in the electrical power generator. The system then reinvested that electricity to power the process, with a 40 percent profit left over to support the colony and heat the buildings. The fresh air system mixed oxygen from the electrolysis system with nitrogen to provide a breathable indoor atmosphere for the colonists.

The system mostly used century-old technology. Titan colonization harnessed hydrocarbon energy, allowing human beings to continue with the same power source they had used to mess up their own planet. But if the tech was old, assembling it in the outer solar system with robots was an impressive achievement. The tinkerer complimented the roboticist, as if she had done it by herself. They spent a lot of time working together and had many such inside jokes, now that they had hooked up.

But as the tinkerer tuned up the system he noticed it was vibrating more and more. Outside, the aluminum front steps of the building didn't touch the ground anymore. On Titan, with its low gravity and cold, dense atmosphere, heated interior spaces had enormous buoyancy. If not securely screwed into the ice below, they could come loose and shoot up into the sky. The warmth of the building had transferred down to the foundation piles and was weakening the ice holding the building to the ground. The astronauts deployed robots with ice drills to install additional anchors holding the structures to the surface. Extra insulation would keep the warmth of the buildings out of the frozen ground so the anchors would hold.

The roboticist created a software interface allowing the astronauts to ride wheeled construction robots around Titan, directing them by voice. Apollo astronauts discovered that getting around the Moon was much easier with their rover, on wheels, rather than with their slow-motion gait as they walked or leaped. The Titan colonists soon treated the robots like horses in the Old West, leaving them waiting just outside the door.

Titan's atmosphere and gravity also made it possible to fly. The tinkerer messed with a pair of wings made of fabric, like a paraglider, and could get aloft from flat ground. But it was slow, as slow as walking. With a thruster—nothing more than an electric motor to blow air, like a reversed shop vac—the wings allowed him to soar like Superman. Even while walking on foot he could go much faster with the thruster helping him along.

After they had refined the system, all the astronauts used thrusters to fly, especially when going longer distances that were inefficient for riding the robots. The risk was low. Flying high in Titan's hazy orange sky, a dead battery or a mistake controlling the paraglider would not be fatal. Even without power, astronauts could glide huge distances before landing.

The fun of flying almost as free as a bird helped alleviate the gloom of isolation on Titan. On Earth, people in developed countries no longer went outdoors much due to concerns about radiation from military conflicts, the heat and storms of climate change, and fear of criminality in disintegrating communities. They stayed inside on their computers. Although the air of Titan was cold enough to kill quickly without a thick heated suit, it felt surprisingly free, deep, and expansive. They could fly and wander in new places without worrying about radioactivity, weather, or hostile human beings, without limits.

But the commander, increasingly alienated from his colleagues, dropped into a dark and uncommunicative mood. His sleep cycle reversed from the rest of the crew. He spent more and more time exercising and he was growing gaunt. He didn't seem to be eating. The physician/psychiatrist tried to recommend counseling, light therapy, and antidepressants, but the commander rebuffed her. He disappeared for long periods.

Finding gear missing from the habitat, the doctor suspected the commander might be thinking of going home. She found him loading supplies into the landing module as if he planned to leave for Earth. As she tried to get him to return to the habitat, he silently refused and lifted off with his fabric wings. As she called for him to return, he flew off over the methane depths of Kraken Mare.

The robot balloons began searching for the commander right away. They never stopped looking, but they never found him. The low density and extreme cold of the liquid methane sea probably meant he was perfectly preserved, frozen solid at the bottom.

In the commander's quarters, the crew found an artificial intelligence companion he had been using, still up on the screen. The others had ridiculed and abandoned these fake friends, but apparently the commander had bonded with one. It had been running continuously for many weeks. The face, a young man, kept asking after him, expressing extravagant praise and admiration. It was love, or hero worship. The crew turned off the application without breaking the news to it.

Earth had preloaded the habitat computer with a training video on holding a memorial service. The astronauts, busy with work and demoralized by the commander's apparent suicide, decided to use the video itself as the service rather than learn how to produce their own. They bowed their heads in front of the screen as the instructors, recorded on Earth, so far away and seemingly so long ago, went through the prayers and ceremonies of a mock funeral and burial.

The remaining colonists had reduced their exercise schedules to roughly what they used to do before leaving Earth. None of them expected to ever return. If the strength of their muscles and bones declined to match Titan's gravity, they could accept that. They did still spend weekly time in a spinning chair that subjected their bodies to enough artificial gravity to circulate their cerebrospinal fluid and protect their optic nerves from swelling. No one enjoyed that routine, but they expected it would be a permanent requirement of life on Titan to avoid blindness or neurological damage.

Their days were full. With the help of the robots, they were building a second and much larger habitat for the next group of colonists, who were already on their way from Earth. They had plenty of time to finish, but the desire to see new faces and receive new supplies drove them to work long hours and to make the new structure as good as it could possibly be. They devised embellishments to the plan—decorations, comfy rest areas, a game room—that would make the newcomers feel welcome and at home. With new people in the colony, the gloom of the suicide and the orange sky would be easier to take.

The doctor had saved a precious box of chocolates and now placed one each on the pillows of the prospective newcomers. During the weekly game night—without cards, they now played marbles with plastic ball bearings from the robot shop—someone pointed out that they had been without chocolate and similar treats for a lot longer than the reinforcements. They debated raiding the pillows for the chocolate. But the doctor said Titan newbies would need it more. Earth would be fresher in their minds and the cold of Titan's nitrogen and methane atmosphere more shocking.

They talked constantly about food. The next spaceship would bring the first new food they had tasted in years. The colonists had finished what they brought with them and were eating from the supplies that had left Earth more than a decade back with the robots, on the slow rockets that had carried supplies for the buildings and machines.

They could not grow their own food. Everyone diligently tended their only plant, the azalea, checking on it every day as if it were a child. That plant, and the indefatigable little flies it still hosted, were their only link to nature. To feed themselves they would need vastly more heated space, more energy, and plants that could efficiently convert electric light into sustenance. Building a system like that would take a lot more people and technology than five lonely colonists could manage.

PRESENT

Producing food in an enclosed system like a habitat on another planet has been tried, and it didn't work out well. In the 1980s a billionaire, Edward Bass, funded an experiment called Biosphere 2, a complete ecosystem isolated from the rest of the Earth (a planet the organizers called Biosphere 1). In the 1970s, science fiction was full of ecology—the science of how energy and nutrients transfer through living organisms—and the prospect of off-Earth colonies or long-range space explorers needing to grow their own food didn't seem as far-fetched with the Moon landing in the recent past.

The glass dome and spacey white pods of Biosphere 2 still stand in the desert north of Tucson. A tour is well worth the twenty-dollar ticket price. Plants grow in a variety of biomes in different rooms—a lush rain forest, a mangrove swamp, an ocean with coral and fish, a savannah, and a farming area for the residents to grow their food. Now owned by the University of Arizona, it feels like an indoor botanical garden. Except for some odd parts. Because the Biosphere was sealed from the outside, it has an underground chamber the size of a gymnasium where a flexible diaphragm would expand to account for the daytime warming and expansion of the air inside the building.

The place was designed to be a human terrarium. It would prove the practicality of living on other planets by creating a complete eco-system meeting all the needs of the eight residents, called Biospheri-ans. While they raised their own food, the plants would emit oxygen for them to breathe, recycling the carbon dioxide exhaled by humans and their livestock. After two years, the Biospherians would emerge as if they had been fully supported by a created ecosystem off the Earth.

Taber MacCallum joined the group eight years before the exper-iment began, while on a three-year sailing trip, an experience orga-nizers expected would make him a good candidate for two years in a dome. That was one of the project's first mistakes. Taber said time on the boat was nowhere near tough enough to prepare for two years of isolation. The crew never did practice with isolation longer than a couple of months. They had no idea what they were getting into.

Once the project started, in 1991, the Biospherians ran into all the problems seen in Antarctica and other isolated environments. They broke into cliques, became depressed, irritable, and conspiratorial, and formed political alignments related to the controllers outside, as loyalists or dissenters.

Mismanagement made it worse. As problems cropped up, including a lack of oxygen inside, managers became secretive, concealing information from their own Science Advisory Committee, which ultimately resigned. The media had lavished coverage on Biosphere 2 from long before it started. When news leaked that early in the experiment the doors had been quietly opened to let air in, without any announcement, reporters turned on the project as a fraud.

The Biosphere 2 project always looked a bit strange and cult-like. The Biospherian lingo and matching suits didn't help, nor their futuristic accommodations (which now look laughably retro). The publicity added an aura of boosterism to the project, a sense that it was a stunt to prove something rather than an experiment to learn something. But for the people inside, it became a long, hard slog to get to the end. And their hardships did provide a lasting lesson.

Besides the stress of isolation, they were starving and couldn't breathe. The crew started on a low-fat, low-calorie diet before they even entered. It didn't provide enough energy for subsistence farming, working all day in their under-glass plots. And crops failed, due to pests and other problems. Without being able to harvest enough protein or calories, crew members became hungry and irritable and constantly fantasized about food. Jane Poynter, in her book on the experience, *The Human Experiment: Two Years and Twenty Minutes Inside Biosphere 2*, said that Taber went from being a bear of a man to being gaunt, as he lost 60 pounds (Jane and Taber got together as a couple before entering the dome and are still married).

"I'd get vivid flashbacks," Taber said. "I would be harvesting peanuts and suddenly I'm six having an argument with my mother. Half of my brain was really in the Biosphere harvesting peanuts, and half of my brain was reliving this classic childhood trauma. Why was my brain doing this? I don't know."

He believes much of the psychological stress came from being confined with just a few people. He sought a solution, not available to space explorers, by calling around to therapists, telling them, "Hi,

I am in the Biosphere and I would like some help." Having telephone sessions a couple of times a week helped.

But it wasn't possible to isolate the variables in this experiment. Hunger could also have contributed to the psychological problems. The crew also didn't have enough oxygen, a situation severe enough to cause sleep apnea. At some points, they couldn't complete a sentence without taking a breath. Conflict became constant. Members of the two cliques didn't eat together or even look one another in the eye. The place was overrun by cockroaches, which had been included to reduce leaf litter but got out of hand. Crops failed repeatedly. In her book, Poynter says all these pressures combined to drive the crew to depression and strange psychological phenomena like Taber's flashbacks.

After the original publicity debacle of letting in air at the beginning, two more injections of oxygen happened later in the experiment. In the final year, people began smuggling alcohol and food to the Biospherians—mostly just treats—but the failure of Biosphere 2 had already been established in the eyes of the public and scientific world long before that. The crew emerged and encountered the vividness of the odors outside the dome two years after they had gone inside, with their dream of scientific credibility gone.

More than twenty years later, tour guides at the site and a film shown to visitors remain defensive. Biosphere 2 advocates point out that carbon dioxide emissions from curing concrete in the building could have thrown off the atmospheric balance, as well as underestimates of how much oxygen microbes in the soil would consume. They seem still to be trying to prove it could work. Ironically, that nonscientific attitude crippled the project in the first place, as managers sought to hide the setbacks. It still obscures a more important scientific discovery: Biosphere 2 did not work.

Biosphere 2 demonstrated that even a huge, carefully designed, richly planted ecosystem under glass cannot feed eight people. The system had all the sunlight in Arizona; all the electricity it could use from the power grid to run water pumps, air-conditioning, lighting, and telephones; and an enormous mass of plants, soil, glass, steel, machinery, water, and much else, far more than could be lifted off the Earth by any technology on the horizon. Yet the inhabitants still struggled to survive on a diet that was more than half fast-growing yams.

The truth is that each human being on Earth is supported by the photosynthesis occurring on a vast area of land and ocean, where sunlight striking plants and algae produces food and transforms carbon dioxide to oxygen. We're overdrawing these accounts, and carbon dioxide in the atmosphere is rising. Crops of food and fiber are squeezing out natural ecosystems as we run out of land. The Biosphere 2 experiment made these trends visible. Eight people need a much larger area than could be enclosed in that dome.

This fact raises the question of how we would feed ourselves on another planet. Fundamentally, photosynthesis is too inefficient. Most of the energy that hits a leaf is lost rather than being captured and stored as food that a person or animal could harvest and consume. Scientists debate whether corn-based ethanol even breaks even, producing more energy than it takes farmers to grow it. The conversion of the Sun's light to biomass is most often estimated at under 1 percent.

On Earth, the ecological system functions because the planet is large and the Sun delivers a lot of energy to its surface. In the outer solar system, sunlight is much weaker; on Titan, under that thick atmosphere, the Sun is always dim. (Even Mars, the next planet away from the Sun, gets 50 percent less solar energy for a given area than the Earth.) Out there, plants would have to grow under artificial lights, and efficiency becomes even more important.

The psychological story from Biosphere 2 is helpful, too. Christian Otto, the physician who studied reactions to Antarctica, also talked to Jane and Taber. They experienced symptoms like post-traumatic stress disorder for some time after emerging from the experiment. Astronauts on long missions will have to avoid their mistakes.

But the couple still dreams of space. Two decades after Biosphere 2, another billionaire was looking for volunteers for a space project. This time, Dennis Tito wanted to send an older couple around Mars (older because they would lose fewer years to radiation), with a small, lightweight spacecraft. A rare period when the planets are especially close was approaching, in 2018. Jane and Taber were chosen. Again they occupied the center ring of a media circus. They worried about the trip. Taber said they lost sleep at night, thinking about what it would be like to be confined together for so long, flying through space.

Part of the concept, however, was that NASA would cooperate with the project. But NASA has other ways of choosing missions and astronauts.

9

WHO GETS TO GO?

It's hard to dislike an astronaut. Yes, they carry themselves as if they think they're perfect. But that's not conceit, it is an accurate self-assessment. They are close to perfect: brilliant, accomplished, well spoken, widely experienced, self-sacrificing team members who also know how to lead. You could try to hate them for being perfect, like a room full of merit scholars and multi-letter athletes in high school. But that doesn't work, either, because NASA weeds out the people whose perfection is grating.

"What you really look for is a nice person," said Duane Ross, who ran the astronaut selection and training program at the Johnson Space Center (he retired since our conversation). "It's got to be somebody you can get along with. We can train folks to do a lot of things. They've got to come with some basic skills, and of course they all come with all varying levels of basic skills. But you need people who can get along with different kinds and types and nationalities of people and do it for a long time in a confined space."

Duane himself isn't astronaut material—he jokes he would scream all the way to space—but he seems to be an excellent judge of character. He came to the job from the oil industry, going from managing personnel who work in the Texas oil patch to a similar job for people who work in outer space. His slow, deep Texas voice places him as an unthreatening regular guy, and it's easy to imagine his career as a long string of job interview subjects being put at ease.

Duane believes in intuition. He came to the job in 1978 to recruit the first crews for the space shuttle. These would be the first civilian astronauts, including women and minorities. For the first time, NASA would look for a range of qualities, not just the fighter pilot's "right stuff." After initial screening for medical and psychological issues, interviewers scored the candidates using a matrix with numeric scores for various attributes. It didn't work. Everyone came out the same. To make a choice, the committee added an extra factor, based on their sense of who would be the best astronauts. That intuitive call proved to be the only one that mattered, and since then the gut has been the main standard.

In 2016, 18,300 people applied to be astronauts, starting an eighteen-month process. The previous selection, in 2013, brought more than 6,000 applicants for 8 spaces (the total astronaut corps has about 42 active members, with 6 flying to the ISS each year). An initial screening narrowed the field to about 4,500 applications. Members of a rating panel—managers and experienced astronauts— were assigned to go through the pile and throw out 90 percent. The 480 highly qualified candidates who survived underwent a review of medical histories and reference checks, narrowing the field to 120 who were invited for preliminary interviews, psychological screening, and medical testing. Fifty made it to the level of a final interview, intensive medical and psychological testing, and a hands-on test replicating doing a job on a spacewalk. Duane made sure their families understood what they were getting into. Then the board talked it over. Anyone could veto a candidate. And they just picked.

So is this arbitrary? Duane says that all the candidates in the final rounds are highly qualified. NASA is like a top Ivy League college choosing among thousands of straight-A students, confident any of the finalists could hack it if chosen. And the system seems to work. It's hard to disagree with the agreeableness of the astronaut corps. Niceness has been Duane's goal, and these are likable people.

He also pointed out that all but a few of the astronauts selected by the program have flown and they have performed well. Looking back over almost forty years of picking astronauts, he thinks intuition has a good track record. There was the female astronaut who in 2007 pepper-sprayed a romantic rival in an airport parking garage after driving halfway across the country in adult diapers. Ross waves that off, saying, "Anybody can nut up." Since that incident, NASA does more psychological checkups after astronauts are working. He said the most severe psychological issues on missions have been disagreements.

The space program began as a front in the Cold War, and missions were manned by military jet pilots, graduates of test pilot school with undergraduate degrees in engineering or similar credentials, who were always under five feet eleven inches tall, so as to fit into their seats. They behaved like military officers, with discipline and a chain of command. Even since bringing in civilians in 1978, NASA still makes it clear who is in charge on every mission and most astronauts come from the military. On the ISS, command switches back and forth between a member of the U.S. and Russian crews.

The Russians' selection process also evolved after the end of the Cold War. No longer did cosmonauts have to be members of the Communist Party or under five feet seven inches tall. But today the pay is poor and young Russians are not flocking to the space program, despite the promise of a luxurious rehabilitation trip to the Canary Islands after each flight. (U.S. astronauts typically start at around $100,000 a year and max out at $156,000.) Chinese rockets have carried five taikonauts into space, including one woman. Female taikonauts must be married, have had a child, have good teeth, and not have body odor. Evidently it's OK if the males stink.

Having power or money also can get you into space. Early in the space shuttle program, U.S. senator Jake Garn and Saudi Arabian prince Sultan bin Salman bin Abdulaziz Al-Saud were payload specialists on shuttle flights, with few real duties. Senator Garn was so sick during his mission that an in-flight sickness measurement scale was named in his honor. (Ross said experience with motion sickness on Earth doesn't seem to predict how the stomach will respond in space.) The Saudi, sent by his father to take pictures of Saudi Arabia, needed to know the direction of Mecca for daily prayers. A company called Space Adventures has sent a dozen paying passengers to the

International Space Station on Russia's Soyuz rockets. Currently the price is $50 million each.

Amanda tried out to be an astronaut once. She made it to near the end of the process, visiting Johnson Space Center for a week of interviews, along with mental and physical testing. Many candidates have to go back several times to get picked. Amanda ultimately followed a different path, with a great career in planetary science, which is lucky, since the space shuttle program was canceled soon after, and she might not have flown for many years.

She recalls NASA's medical tests looking for people who could fit the suits and equipment, and whose bodies wouldn't break far away from the nearest medical facility able to fix them. Ross said the tests are designed to find out if an astronaut will have health problems in ten years, since it is likely a novice won't make it to space any sooner. Amanda's elbows and knees were tested for strength, she had vision and hearing tests, ran on a treadmill with various sensors attached to her body, sat in a dark ball to see if she would stay calm, endured an enema and a colonoscopy, a psychological interview, and ate dinners with other candidates, during which their behavior was observed. Since then, the program has added MRIs to look for potential brain aneurysms and an ultrasound for susceptibility to kidney stones, a frequent problem for astronauts in weightlessness.

Duane believes in finding the best people and then preparing them for whatever they will face. Every ISS astronaut has to learn to fly a supersonic T-38 jet trainer and to speak Russian, in case of an emergency on the Soyuz spacecraft. They work underwater in the world's largest swimming pool, 12 meters (40 feet) deep, to get weightless experience handling whatever equipment they will use in space. Virtual reality helps get the details, too. Basic training takes two years and mission training takes another two to three years.

Duane would use the same process for a crew going to another planet for a long mission. The basic selection system works. The psychological screening process already gives each astronaut different ratings for short and long missions. New psychological screening might be able to choose people who handle very long isolation and confinement well—if research can find the markers for those qualities.

Duane thinks crews of two or three wouldn't work, as the consequences of conflict would be too high. Four or six could make sense. And then you would need to decide if the astronauts should be single, married, or married couples. Romantic relationships probably do occur in space. Even with the brief time available on shuttle missions, rumor has it hookups did occur, as we noted in chapter 7. ISS missions last six months, astronauts have privacy, and NASA doesn't ask questions. The problem Duane worries about is not hookups, but breakups. How would a long mission work with an ex on board?

No matter how crew selection for a trip to another planet works out—and it's still in the realm of cutting-edge research—Duane is certain the mission will go better if the astronauts are nice people. And that's best decided by the gut feeling of the selection committee.

FUTURE

The world had watched with rapt attention as the first colonists landed on Titan, set up their home, lost their commander, and prepared for the next group to come, a party of six more astronauts on a sister ship to the one they had arrived on. After building the first two Q-drive spacecraft, Titan Corp. won a contract for eight more, five to carry food and other time-sensitive cargo and three to carry additional crews of six each. That would give Titan twenty-nine residents in three habitats.

Slower, heavier freighters were commissioned, too, to bring cargo for a massive expansion of the colony. Engineers packed parts for a plant to produce plastic construction materials from the hydrocarbons on Titan's surface, which would allow the colonists to erect large buildings from the materials at hand.

Events on Titan began to eclipse the news of the disasters on Earth. The colonists' status exceeded that of any other media stars. To viewers on Earth, they felt like family members, their stories playing out on screens everywhere (at least everywhere on Earth that still had electricity and connectivity). The soap opera from outer space allowed people to stop thinking about their dysfunctional governments, stratified societies, endless wars, climate crises, scary robots, and other earthly troubles. It was easier to throw up your hands and look to the sky. On Titan, they could see people building something new, unafraid of violence or radiation, able to fly through the air and explore brand-new horizons. The hardships, cold, and darkness didn't come through as clearly on the screen.

A group of world leaders from Western powers contracted with Titan Corp. for ten spacecraft, each large enough to carry one hundred colonists. The first colonial liner would be named the *Mayflower*. The deal was exclusive, and its announcement forced other wealthy nations to come on board, even if they were generally hostile to the United States and Europe, which were leading the program. Catching up to Titan Corp. would be too difficult and expensive even for China, stressed as it was, like other nations, by droughts, rising sea levels, violent dissent, and a broken international trade system.

The public had accepted an opaque selection system for the initial crews going to Titan. Obviously, only the most capable astronauts could be sent on the five small initial missions. But the thousand seats to fly on the ten big ships—those would be tickets off the Earth, the chance to start again in a place where it was safe to go outside (if cold). On Titan, families could survive for future generations and religious and cultural traditions could live on indefinitely, as they might not on Earth if its societies and biosphere were to collapse totally.

Political leaders could see they had a problem. To keep the coalition of nations together, and avoid domestic political upheaval, they needed a selection program that would look completely logical, fair, and transparent.

A blue-ribbon commission convened to select the colonists. With many high-flown pronouncements, the international partners agreed that the commission should include the best minds, representing all facets of human endeavor. These authorities would set the standards and decide the process for choosing a perfect cross section of humanity, people who could become the progenitors for all future off-Earth generations. The historic nature of the task—to debate the essential qualities of our species—called for a diverse and talented group, expert in everything.

The commission's membership was large, with top scientists, physicians, psychologists, ethicists, religious leaders, engineers, teachers, and military leaders; there were professors in painting and sculpture,

literature, music, dance, video game design, and the new art of creative biology; as well as specialists in anthropology, sociology, and
gender studies and a wedding planner (this last a person who was
familiar with managing large, diverse groups of people in high-stress
situations).

The first all-day meeting convened. After a facilitator led the
commission members through a team-building exercise, a brainstorming session, and an exercise writing notes on yellow stickies and
putting them on the walls around the meeting room, discussion could
begin. But the preliminaries had taken up most of the time, and the
open session was dominated by a long, testy discussion by two self-
important members of the commission about the name of the first
ship. They argued about whether the historical echoes of the name
Mayflower suggested escape from an oppressive Old World religious
system or represented white, European hegemony and genocide of
indigenous people. The group adjourned inconclusively and a staff
member went off to write the report on colonist selection based on
the research that had already been compiled before the meeting.

The colonists on their one-way trip needed qualities besides the
ability to work in a team. Those chosen might be the future of our
species, so the commission's staff wanted each to be the best of his
or her type, as well as young, fit, healthy, and suitable for breeding. Besides medical screening that eliminated any potential hereditary flaw or likelihood for illness, potential colonists would have had
genetic analysis for traits that would improve their chances on Titan.
Genetic markers would disqualify applicants with increased chances
of elevated eye or brain pressure in low gravity, depression in low
light or confined conditions, experience of cold extremities, or an
aggressive temperament.

Staff members drafted qualifications for all one thousand slots,
each assigned a collection of attributes. Obviously we would need
nonfiction writers and planetary scientists—at least one of each—
along with many other professionals, while simultaneously keeping
the gender balance and being certain to cover varied races, cultures,
religions, and political outlooks, and including straight, gay, bi, trans-
gender, and a variety of newer sexual identities.

The goal of diversity and political correctness sometimes collided with the goal of compatibility. Representing many religions would be easy enough, but bringing a cross section of faiths would mean sending a good many fundamentalists of various religions who might have a hard time getting along with one another. How to handle those who believed that the Earth was at the center of the universe? The spaceship couldn't be made to obey Sabbath rules for Orthodox Jews; candles wouldn't burn properly in zero gravity for Roman Catholics; and Rastafarians could neither grow nor smoke their ganja in space. The commission's executive director drew the line when presented with a proposal to send Muslim extremists who might want to blow up the spaceship.

The complex selection matrix required a long and personally intrusive application. People looking for a free ride off the planet filled out online forms that asked for their most intimate medical, social, and sexual information, as well as background on their work, awards, recommendations, and personal essays. It was like a college application on steroids. The concept of privacy had long since become obsolete, so many applicants already stored the necessary information on Facebook-like profiles that auto-filled the form.

The commission's servers gathered millions of applications, sorted and organized them, swept the web for additional information on each applicant, noting inconsistencies, and scored them according to a rubric based on the experts' prior opinion of what a perfect colonist in each category would look like. The computer delivered a name and an alternate for each of the thousand slots, having matched the preset criteria across all the possible permutations. The commission studied the list, then retreated into a secret meeting for a hot session of bargaining and trading that led to overruling some choices in favor of candidates who knew how to pull strings.

Despite any compromises, however, the thousand colonists truly looked like model humans—beautiful, accomplished, well-adjusted, healthy, and of every color (although the mix was heavy on white and Asian faces from the nations paying for the trip). They became local celebrities in their own communities, the chosen ones.

But as training for departure geared up and construction began of an immense space dock to build the first, now unnamed ship, the backlash began. At first, the Twitter comments were snarky. Who would want to live on a planet of perfect people anyway? It would be like a high school full of teacher's pets. Then the more serious criticisms. Wasn't this the kind of selection the Nazis and the other eugenicists of the twentieth century had in mind when they sought to weed out the undesirable elements of humanity and allow only those they considered the best to breed? Did we believe people with disabilities should be eliminated from our species? What of the mad artists whose genius goes unrecognized in their lifetimes? There would be no bipolar Vincent van Gogh on the spaceship, no learning-disabled Agatha Christie, no wheelchair-bound Stephen Hawking.

And from this first grumbling, the roar of objections exploded, augmented by powerful people whose consultants knew how to amplify their voices on social media. They said the successful should go—the strong and aggressive who had won in life. They often were not the ones who got the best grades, pleased academic committees, or won awards; they didn't work the system, they beat it. They were the ones with egos, inventiveness, and daring. The survival of the fittest elevated the rich and powerful, and they were the best chance, so went the argument, to make sure our species survived and thrived in the future.

And they should also be allowed to bring along their pets, wine collections, and fine art.

The original *Mayflower* of 1620 was not an ark, built to carry the best examples of each type of person to new ground in which to plant a civilization. Its passengers were escaping a home country that had become unlivable for them. They were the most committed, the ones brave enough to make a new world, not darlings of the establishment but self-selected escapees. The space colony should be a lifeboat for those willing to fight for a new world.

Or so the critics said. They would have said anything to get a seat on the spaceship.

PRESENT

Probes exploring for the site for a new colony may send back a misleading impression. In 1584, explorers sent by Sir Walter Raleigh landed on the Outer Banks of North Carolina and found a paradise where food grew profusely without effort, natives were friendly, and prospects for finding precious metals appeared good. The ships were self-contained. They needed little from the new land to survive. Their captains returned to England with an impression like that of a tourist, who gathers the highlights of a new place without a sense of the real difficulty of living there. Planetary probes of Mars and Titan also provide pictures that look something like Earth, but staying there self-sufficiently would be infinitely harder than going for a visit.

The differences between space colonization and the early colonies of North America are obvious—for one thing, we won't run into beings similar to us anywhere in our solar system—but the similarities are striking, too. The English view of the American coast in 1584 wasn't so different from our view of the planets. Ships had sailed, landings had been made, but details were scarce and discoveries from close contact had not yet occurred. Sending ships to America was expensive and risky, and long periods elapsed between successful voyages. And nations competed. Discovery for science and prestige went along with simple greed and hope for new opportunities.

The fearsome distances and hazards were similar, when scaled to the speed of the craft. Crossing the Atlantic in the sixteenth century typically took a couple of months and sailors commonly left home for years. Communication relied on letters carried by luck to the right destination by other ships met by chance. The attrition of sailors far exceeded that of astronauts. If only a few died, that was a good voyage. Officers commanding long voyages also endured psychological hazards like modern workers in confined postings such as Antarctica or Biosphere 2. Intense, irrational conflicts color the journals of voyages involving colonists, scientists, or others outside the military chain of command, and sometimes also among those who risked summary execution for talking back.

If human nature has remained constant over the last five hundred years, space colonists will be driven by the same social and economic forces, will make the same mistakes and miscalculations, and will face the same authentic risk. A colony might make it on the first try, and it might not. A lot of people could die. The first pioneers may pay a high price. But eventually, colonies take hold.

Elizabeth I pioneered an economic model for colonization that Americans used for another three hundred years, through the conquest of the American West, and continue to use on technological frontiers: a public-private partnership. The dashing and brilliant Raleigh was her favorite courtier. She gave him a charter to claim ownership of North America if he could settle people there—just as Congress later gave vast western lands to railroads as they crossed the continent and gave broadcast spectrum to corporations building radio and TV stations.

While Shakespeare wrote his first plays and the creativity of the Renaissance bloomed in London, England remained a weak country threatened by the overwhelming wealth and power that Spain had gained from the gold it took from America. Elizabeth couldn't afford to take on Spain directly, but she subsidized construction of ships for merchants that could be converted to effective fighting vessels if military service was needed. And she encouraged privateers, essentially legal pirates, to attack Spanish ships in the Caribbean and give her a share of their winnings. Her deal with Raleigh promised a similar shared investment and reward. A successful colony would increase his wealth and expand her power.

Raleigh was a venture capitalist as daring as any Silicon Valley visionary, playing with high costs, high risks, and the possibility of an immense reward. Spain hadn't settled north of Florida. A colony up the coast would help secure that territory for England, give rich discoveries and enormous landholdings to Raleigh, and provide support to English vessels raiding Spanish treasure ships. But even the geography of the area was unknown. Raleigh sent his pair of ships in 1584 to probe the coast in search of a harbor and land for settlement, like planetary craft on a flyby.

The barrier islands of the Outer Banks line the coast of the Carolinas like a sandy parenthesis. Within this long, narrow strand lie broad sounds, many miles wide, that are too shallow for large vessels. The changeable inlets that cross the Outer Banks are tricky and impossible for large ships to navigate, and the waters outside so hazardous they became known as the Graveyard of the Atlantic. Landward of the sounds, much of the ground is low and swampy. Neither the sandy islands nor the wet mainland offer much promise for growing crops. Roanoke Island, within the sounds, is relatively protected and mostly dry ground but not large enough to support many people. This was not a good place to plant Europeans in America.

Unfortunately, Raleigh's first expedition didn't go a little farther north to find Chesapeake Bay, with its superior land for settlement and ideal protection for ships. Raleigh instead received a report of paradise at Roanoke Island. Two Native Americans, Manteo and Wanchese, came back to England with the expedition, creating great excitement for the voyage Raleigh planned to send next. This voyage would be much larger, with a mix of people selected, much as we would imagine for a planetary outpost. It would seek a solid understanding of the new land's potential for mining and agriculture, in-depth scientific exploration, and creation of a permanent town site for colonists who would follow.

Queen Elizabeth knighted Raleigh and contributed to the expedition with gunpowder and a ship and gave him the title of governor of Virginia (named after her, the virgin queen), lands including imaginary control of 2,900 kilometers (1,800 miles) of American coastline and the balance of North America to the west. And she gave him the right to conscript men to join the voyage. London was crowded, jobs and land were scarce, and America sounded like a great opportunity. Raleigh wanted to go, too, but Elizabeth wouldn't let him because she wanted him at home.

Raleigh hired experts to advise him on the kinds of people to send on the second expedition. There would be soldiers in a role like our astronauts and for defense from the Native Americans, scientists to gather information, businessmen paying and seeking their own profit, experts on metals, gems, and mining, botanists and experts on medicinal plants, specialists who could support the rest by raising, processing, and storing food, forging tools, weaving clothing, and cobbling shoes, and tending to the sick and injured, and spots for many other professions, and a hierarchy of officials to manage them. The list ran to more than eight hundred people and kept growing, as David Beers Quinn reports in his excellent book *Set Fair for Roanoke*. Raleigh cut it to five hundred to travel on various ships in two successive fleets.

One of the most important lessons of the colony project, however, was that plans don't go as expected. Ships sank or went to the wrong places and various other mishaps intervened. Roanoke Island was too small to support a large group. The expedition ended up leaving only about one hundred men there, at least half of them soldiers. Ships left that outpost behind with the promise of resupply in a year. But the circumstances left the colonists without enough supplies to last that long. To make it, they counted on unrealistic estimates of their ability to produce food or get it from the Native Americans.

Raleigh's best choice for the expedition was Thomas Harriot, a brilliant scientist and mathematician who had helped improve navigation and may have gone on the first voyage as well. John White, an artist, handled the imaging that is still such a critical part of exploration. He also proved to be a perceptive observer whose work would capture the nature and people of North America in a vivid way for the first time for English viewers. He returned with the ship that dropped off the one hundred men with watercolors and positive reports.

Harriot stayed the entire year. He had learned the native language from Manteo. He trekked far and wide, visiting indigenous villages and capturing information about their cultures at a unique time, the dawn of contact, a moment that could never be replicated. The information, now a treasure, became a state secret, with White's watercolors, as the basis for building a successful colony.

"There was a lot of interest at the time in pharmaceuticals," says historian and archeologist Eric Klingelhofer, of Mercer University. "They thought sassafras was a miracle cure, so anybody who could find a tree and get some roots of sassafras could make some serious money in Europe at the time. And they thought other things had a good chance. A lot of Harriot's book is just looking at these different plants and saying this could be applied for this and that, the Indians say it is good for some stomach ailment or whatever."

But Harriot had contempt for the wealth seekers on the expedition, who had paid for their passage in hopes of finding gold or rich trade with the Native Americans and, finding none, spent their time complaining about the food and accommodations and doing nothing to help. The later Jamestown colony in Virginia ran into the same problem with gentlemen who didn't work while the entire party almost starved.

"They had little care for any other thing but to pamper their bellies," Harriot wrote of Roanoke Island's wealth seekers and complainers. "Because there were not to be found any English cities, nor such fair houses, nor at their own wish any of their old accustomed dainty food, nor any soft beds of down or feathers, the country was to them miserable."

The ecology of a functioning human community is shaped like a pyramid, with a broad base of producers and a narrow space at the top for idle consumers. Contrary to the evidence of BBC costume dramas, most people in England were not lords and ladies but peasants who farmed crops, spun fiber, and carried water. Only a tiny fraction of the population could afford to sit in manor houses and do anagrams. A truly self-sufficient colony of one hundred might not be able to afford a single person who didn't pull his or her own weight.

That lesson is still true. Automation and powered equipment certainly improve the ratio of leisure to human toil in our society, but living on another planet will require much more fundamental work just to survive. Sixteenth-century communities grew crops to feed cows to produce milk to make into butter and cheese, with skilled workers all along the chain of production. Imagining the same system on Titan (with ample poetic license), we would need additional links in the chain: people to build a habitat for the crops and animals and to provide warmth, light, and water for life and oxygen for the cows to breathe. We could give up naturally produced dairy products, robots and other machines could do a lot of food production, and biotech could make it more efficient, but we've lived with technology long enough to know that it doesn't eliminate human work. Someone has to operate the machines.

At Roanoke, the strength of the military contingent took the place of food production. Rafe Lane, the military commander, first traded with and then pressed and threatened the Native Americans for food. Wingina, the chief on Roanoke who originally befriended the colonists, eventually withdrew his villagers from the island, away from where Lane could easily intimidate them. His people were probably going hungry trying to support themselves and the colonists. Lane got word that Wingina was planning an attack—and maybe he was—and requested a meeting. But at the meeting Lane ambushed Wingina and killed him and several of his top men.

Harriot also harmed the Native Americans, although unintentionally. The villages he visited suffered deadly epidemics soon after he left. European germs the indigenous people hadn't encountered before struck down huge numbers, probably with influenza, smallpox, and the like. The natives believed Harriot had invisible power to inflict disease and death. Harriot, a man of his times, had a similar theory about the outbreaks, thinking they represented some kind of divine justice for the natives' deceptions or slights.

When a year had passed, the colonists on Roanoke Island were hungry and worried about the Native Americans and the Spanish. They desperately needed supplies and reinforcements. But the ships Raleigh sent were late. When Sir Francis Drake showed up unexpectedly with a strong fleet, fresh from successful attacks against the Spanish, Lane had little choice but to accept a ride home.

"They really were thinking twelve months, and then supply ships will come and replacements will come," Klingelhofer said. "Well, neither came, so they just looked at their watches, or sundials, and said, 'Well, time's up, we've got to get out of here. We've done pretty much what we could, and we haven't lost very many men—no men yet, actually, through warfare—and without supplies it is no good staying.'"

When the resupply ships did arrive, their officers found the abandoned colony and only three men stationed there by Drake. They left behind fifteen more men with two years of supplies. Those men apparently didn't survive long before neighboring tribes joined forces and eliminated them.

The third phase of Raleigh's project went forward with the benefit of knowledge gained by the first two expeditions, including the need to be self-sufficient. The makeup of the colony was completely different and revolutionary for the time. Families would go to America, with women and children, to build farms and stay permanently. Investors other than Raleigh would stake the funds through a corporate structure. And the knowledge of geography gained from a year of study and exploration would put them in an area with a greater chance of success: Chesapeake Bay.

John White, the painter, gathered up about 115 colonists. He tried for more, but recruiting was difficult. He could offer real estate to land-poor Englishmen and a healthier climate with less chance of death by disease than in London, but the risks of the venture were enormous, especially for families who already had some assets and a future at home. Those we know about were middle-class people without titles, but with resources, and their servants. Many may have been Protestants who wanted a freer form of religious worship than the Church of England allowed, but they weren't extreme religious dissenters like the Pilgrims, who came later. They were well prepared and supplied and their plan could have worked.

But White wasn't a good leader and the group had terrible luck. They left much too late in the year to allow planting in America, so they would have to last a year on supplies from home or from the Native Americans. White and his primary ship's captain got into continuous, toxic conflict. They arrived at the Outer Banks, after numerous hardships, having failed to obtain critical supplies of food and livestock during the voyage. On Roanoke Island, they discovered only bones from the eighteen men left behind. And then, for reasons that remain murky, they decided to stay there rather than continuing to Chesapeake Bay.

By the time their ship was ready to leave, it was obvious the colony would need to be resupplied to survive. The colonists insisted on sending White back for help. His daughter had given birth to the first English child in America, Virginia Dare, so perhaps they thought he would be well motivated to return. Or maybe he was such a poor leader they wanted to get rid of him. He sailed off, desperate for help.

But world events overtook the colony. War with Spain heated up. Ships that Raleigh and White had arranged to support the colony were diverted by order of the queen to attack Spanish ships in Newfoundland. Then Spain launched its Armada to invade England, the greatest naval attack in history, and every vessel was needed for defense. We can wonder what would happen to a space colony in a similar situation, if it needed resupply when world war broke out on Earth. Would a president spare great wealth and top technical workers to supply one hundred colonists in the midst of total war? Elizabeth did not. White wasn't able to return to Roanoke Island for three years.

When he arrived, no one was there. The houses had been dismantled and removed. A strong fort had been built and abandoned and was now overgrown. Valuables had been buried and then dug up by the natives. A one-word clue had been left, carved in a log: "Croatoan." White had made an agreement with his son-in-law and the others that if they had to move, they would leave the name of their destination carved in wood. Croatoan was the sandy barrier island where Manteo's still-friendly tribe lived.

Croatoan was less than a day south, at today's Hatteras Island. But a storm whipped up that night and the ships had to escape to sea. They nearly wrecked in the dangerous waters off the Outer Banks. White's ship tried to get back, but Raleigh had sent him with privateers who were really more interested in attacking Spanish ships. They never returned to the Outer Banks. No one looked for the colonists for many years, and no one ever found them. A generation later, green-eyed Croatoan Indians told visitors they were descendants of colonists who had melded into the native culture, and there's no reason not to believe them.

Today, archeologists (including Eric Klingelhofer) continue to search for the lost colony. In 2015, Eric's First Colony Foundation announced finding clues to the colonists' presence at a dig 80 kilometers (50 miles) west of Roanoke Island, on the mainland. Some evidence suggests they went north, toward Chesapeake Bay. It's likely that they split up, with better chances of feeding themselves in smaller groups.

Later colonies survived, including Jamestown in Virginia and Plymouth in Massachusetts. Information was key, and geography, and luck. And, for the first generation, hardship, so that succeeding generations could have easier lives. The same may be true for space colonists.

What's certain is that the selection of who will go will be essential. The colonists must have the skills for self-sufficiency. The lost colony of Roanoke Island taught, above all, that colonists can't trust a lifeline to supply them from home. They need the toughness and resources to survive on their own.

FUTURE

The thousand chosen space colonists trained at an indoor facility in the high desert of Colorado, remote from rising seawaters and breathing air filtered to reduce fallout and pathogens. With the pace of world disasters accelerating, the mission originally conceived as a bold and adventurous outward step took on a new meaning. The colony mission now seemed like the only safe way off a planet that was spinning out of control, toward a new dark age or even human extinction. The thousand began to look at one another as the last human beings, a collection of perfect people in matching orange jumpsuits.

Others thought differently. The colony might be the key to personal survival, or a way to pass on the precious line of family genes to future generations. The superrich could defend themselves on Earth in environmentally contained compounds, insulated from the effects of climate change, natural disasters, and nuclear war, but they could not evade the sense of dying hope on a world that had failed to contain its demons. The future would be on Titan, and so that was where the most powerful humans wanted to go.

Construction of the first large ship for the colonists went slowly; nothing like it had been built before. An accident slowed the work further. A supply ship headed to Titan disappeared, perhaps the victim of a meteoroid strike. To keep the settlers at the Titan outpost from starving, every worker at Titan Corp. diverted to an urgent program to build replacement supply ships as rapidly as possible.

The government consortium paid the bills. Titan Corp. owned the technology and controlled the facilities to build the ships, with government orders on hand for ten spaceliners, each to carry one hundred colonists. But with the delays and uncertainty, the governments paid for only the first three.

A billionaire who knew the CEO of Titan Corp. from their days in an elite prep school pulled her aside at their tennis club. He offered to buy spaceliner number four and pay more than the government had budgeted for it, with money up front. After a quick series of secret meetings, the deal went through in a matter of weeks.

When Titan Corp. announced the private purchase of spaceliner number four, a bidding war erupted for the other six planned liners.

With those sold, Titan Corp. added and sold more ships on a schedule far into the future. The governments, with their cumbersome international structure for making decisions, didn't even convene a meeting to discuss the problem until it was too late. For the foreseeable future, the official Titan colony would have the ability to deliver only three hundred members, to be followed by who knew how many more private passengers of unknown preparation and skills.

The space colony had been planned in one piece as a carefully structured project managed by government bureaucrats. But the new private colonists objected to central management, just as they had resisted the idea of a eugenically screened crew. Under government control, this new world would be utterly communistic, in a way that even the old tyrants of the twentieth century had never achieved. Everything would be owned by the government, everyone's jobs and homes dictated by the government, even decisions about health and reproduction would be managed. And there wouldn't be any way to make money.

Unacceptable!

Lawyers got involved to create opportunities for their clients. Attorneys for an aggrieved billionaire noted that no Earth government could claim ownership of the extraterrestrial colony because of a 1967 treaty that stated, specifically, "Outer space, including the moon and other celestial bodies, is not subject to national appropriation by claim of sovereignty, by means of use or occupation, or by any other means." The government coalition had pioneered the new colony but could not claim to own it or to control who else would go there. The fractured international political system made amending the treaty inconceivable.

But the treaty made no reference to private claims of extraterrestrial ownership (no one had thought that this would be an issue in 1967). Under its terms, a government couldn't own another planet, but maybe a private person could. Around the world, heads of large corporations realized that the first private entities to arrive at the colony could claim vast land ownership and exert great influence in creating the government that would rule there. Visions of space feudalism flashed in the minds of these heretofore only metaphorical masters of the universe. On Earth they could only be CEOs, but in space they could be lords, barons, or kings!

But all the excitement and money couldn't get the ships built faster. Private owners spent their time redesigning and decorating the three-dimensional computer models of the interiors of their future spaceliners. The government's volunteers had expected to fly in Spartan quarters; the private ships were designed to carry fewer passengers, but to carry them in luxury, with servants.

Only the ultrarich could afford to go in that style, but a group of entrepreneurs recognized the opportunity for many more wealthy people to propagate themselves on another world. They designed their ship with liquid nitrogen casks to carry ten thousand embryos, to be staffed by one hundred nubile young women willing to be repeat surrogate mothers in exchange for their passage (tubal ligations guaranteed they couldn't become pregnant with their own genes). The fare to send an embryo depended on your priority in the womb schedule: to be in the first round of births was most expensive, while embryos without any committed implantation time could go relatively cheaply, able to wait decades or centuries to be brought back to life.

Originally the embryo mission had been designed with an all-female crew: doctors, nurses, engineers, and technicians were among the surrogates, staffing an estrogen-fueled baby factory in space. Then the organizers realized they had created a male fantasy: an alien world populated only by beautiful young women. The bidding went through the roof for a few extra seats sold to men.

Space colony leaders from the governments that had started the project met secretly to consider the situation that had developed. They would rework their crew list and equipment to set up a smaller colony than they had originally planned, one that could still be self-sufficient with around three hundred members. But as they monitored the circus of private entities buying their way to Titan, they realized that the government colony couldn't support all those other people. If the private colonists ran out of food or supplies on Titan, they might pressure the government's colonists for help. For the first time, the colonists' training began to include military skills, and weapons were added to the cargo on the first three liners.

Organizers of the private ventures had the same thoughts, but along different lines. What if the surrogate mothers declined to be implanted once they were a billion kilometers away from Earth?

Would the billionaires' servants agree to remain servants when they outnumbered their masters at a remote space base? Without a government, a legal system, or even a real economy, the systems of compliance that keep control on Earth would be mostly absent. The rich wouldn't be rich anymore if the people working for them refused to go along with the fiction of their status.

The first-wave colonists already on Titan struggled to stretch their supplies until a replacement freighter could arrive. Viewers on Earth watched, inspired by their representatives on Titan, who were sacrificing for the future of humanity on a cold, dark world. Meanwhile, the second wave of colonists prepared for conflict and potential warfare.

The stock price shot up in a new company selling guns designed to be fired in Titan's atmosphere.

PRESENT

The vehicle most like a spaceliner that exists today is a U.S. Navy fast attack submarine. It also carries a nuclear reactor and a crew of around one hundred, of both genders, with a range limited primarily by the amount of food that can fit on board. Submarines obtain oxygen the same way the Titan colony would, by splitting it from water with electrolysis. The overall length of a sub is about the same as the International Space Station, although a Virginia-class submarine weighs twenty times more and has much more interior space (but the ISS can go four hundred times faster).

Submarine crews carry immense responsibility. Their competence and psychological stability couldn't be more important. On ballistic missile subs, they lurk under the ocean for seventy-seven days at a stretch without sending messages to the outside world, prepared with missiles that, if fired, would wreak destruction on a scale never before seen on Earth, perhaps dooming civilization. Attack submarines are intended for conventional warfare, but their sailors also operate nuclear reactors as dangerous as any in a power plant.

In more than fifty years, the navy's record of safeguarding these nuclear materials is close to perfect. Engineers say submarine reactors are reliable because they are reproduced many times rather than being built individually like nuclear power plants. But submariners say crew selection and discipline is the key. Nuclear submarine personnel are not known for being easygoing. They do everything by the book. Their culture rests on the assumption that anything that isn't specifically allowed in the navy's regulation manual is prohibited, without compromises.

"The cornerstone of the nuclear power program is integrity," said retired commander Rick Campbell, who spent a twenty-eight-year career in subs and now writes suspense novels. "If you would ever be caught in a single lie, you're gone from the nuclear power program. If you cheat on an exam, you're gone. When you deal with nuclear power, you can't have any doubt about whether someone did some of the maintenance they were supposed to do, about whether they did a procedure correctly or incorrectly. Everything is founded on integrity."

Submarine crews pass intelligence and psychological tests. Rick said psychological problems on board are rare. He never encountered a case of claustrophobia, and the young men he commanded were kept too busy to get depressed, with constant schedules of training and maintenance. A submarine's behavior problems more often relate to youth and masculinity, with a crew of teenagers and guys in their early twenties, the oldest man on board in his thirties. Women were added recently. At sea, the navy's rules about gender are simple: no relationships and no sexual contact.

On ballistic missile subs, sailors' incoming messages from home are screened. Officers don't let them have bad news, like deaths or relationship breakups, to avoid mental disturbances. Campbell created a webpage that advises crewmembers how to prepare for a deployment. He would write a batch of letters to his young daughter for her to open as the months passed.

Fast attack submarines deploy for six months at a time, usually with a few ports of call where they pick up fresh food. Food is the limitation. Modern nuclear subs never need refueling during their twenty-five-year service life and could theoretically stay underwater for decades if they didn't need food and spare parts. In fact, stretching supplies to six months is difficult. Veterans swap stories on Internet message boards of extended voyages when food ran short and they ended up eating nothing but pancakes or weird combinations, like chili mac and canned beets.

Generally the food is good, by military standards, a selling point in navy recruitment videos and a time-tested solution to morale problems. Crews have regular movie and pizza nights and eat plenty of meat and freshly baked bread and pastries. Like Antarctic overwintering scientists, submariners plan a big feast for the halfway mark in a voyage, called halfway night, with a party, games, and relaxed discipline, and a dinner of lobster and prime rib.

A Virginia-class attack submarine is huge, 115 meters (377 feet) long and 10 meters (34 feet) in diameter, but cramped. The crew walks along three levels of decks (or sleds along, when the ship maneuvers up or down), and climbs ladders between decks. The bow of the ship contains navigation, command, and weapons systems. The middle section is crew quarters and dining facilities. The reactor and engine room are in the stern. Each section takes about a third of the space.

Crew members, except for a few top officers, sleep in racks of three bunks, which are called coffins because of their size and shape. Each bunk has a single row of drawers underneath. That's a crew member's entire personal space. The ISS, with only six crew members on board, has plenty of personal space. Unlike astronauts, nuclear submarine crews do get to shower and wash their clothes. Astronauts take sponge baths and wear the same clothes over and over until disposing of them. Modern submarines and the ISS both have effective air handling systems. The most memorable smell on a sub or the space station is a lack of odor that contributes to the overall sensory deprivation of the journey.

The navy knows how to find things, a problem it solved long before the space station began orbiting and ran into its chronic difficulty with lost items. A submarine is packed with spare parts hidden in a multitude of lockers, all tracked with a formal system. Subs have storage space for food lasting two or three months and can extend their voyages by packing food into every possible area. Canned goods cover the decks, topped with rubber mats for walking. As the voyage progresses, the crew eats its way down to the real deck.

A spaceliner with a nuclear reactor might need as much room as a sub for its mechanical and command areas, but it could presumably get by without torpedoes or missiles. However, a craft bound for Titan would need a lot more storage space to feed people for several times longer, and that assumes a supply cache or renewable food source available at the other end of the trip. A spacecraft would also need to carry much more water. Submarines purify the water they need from the ocean. The ISS reuses water. What goes down the toilet comes back around again in the drinking water supply, over and over.

Crew skills would be different, too. Rick pointed out that a submarine carries a medic who is qualified only to stabilize an injured or ill crew member for transport to a hospital. A spaceliner would need a complete medical, dental, and psychological team, and backup personnel in case one of them got sick.

The purpose of the two kinds of craft is different. The purpose of an attack submarine is to sink ships and other submarines, fire cruise missiles to destroy enemies on land, and deliver special forces warriors. In a world without conflict, big submarines wouldn't exist. A spaceliner wouldn't be built to kill, but it also might not exist in a world without conflict. If we're right that the main impetus for space colonization would be fear about the future on Earth, we can also predict that a human species that could work together to solve its problems wouldn't have those fears.

Fear and conflict have motivated investment in spaceflight before. The United States spent the money to get beyond low Earth orbit only during a period of intense Cold War conflict with the Soviet Union, when Americans believed technological superiority might help avoid annihilation. But success in space also inspired millions and renewed their faith in humanity, creating a sense of pride in our species and a new awareness of our place in the universe. Could we build a craft as big as a spaceliner for that purpose instead?

10

WHY MOVE INTO SPACE?

Pioneer 10, the first spacecraft to go to the outer solar system, carried a plaque showing the location of the Earth, a few science facts, and a naked couple, a man and a woman, the man waving his hand in greeting. Carl Sagan got the idea for the plaque from a colleague and suggested it to an official at JPL before the launch in 1972. They quickly obtained a design and had it engraved on aluminum at a local shop and bolted it to the spacecraft without telling their bosses. Carl later explained the purpose: with Pioneer leaving the solar system, someone out there might find it, and the plaque would tell them where we are.

Or where we were. Sagan said it won't get anywhere until every trace of humans living on Earth is gone.

"It will very likely survive for billions of years, but in the dark of interstellar space," he told a BBC interviewer. "It will be the oldest artifact of mankind."

The plaque means something. The knowledge that our image travels onward eternally in space—that knowledge has spiritual power for many people, including the staunch atheist Carl Sagan. But why is the plaque important? If someone finds it somewhere unimaginably far away in time and space (an extremely unlikely event), no practical purpose will be served by the discovery. Instead, the purpose of the plaque is today, for us. The plaque declares, "We are here. We exist." It's cosmic graffiti, a permanent tag naming humanity. The message feeds a hunger for immortality, at least for the species if not for ourselves.

The desire to colonize space comes from the same hunger. At most, an infinitesimal percentage of humanity's total population will make it off the Earth to any colony. But those few will carry genetic information that all of us share, a voluminous library of data compared to the graffiti scrawl of a single anodized aluminum plaque. A space colony means we won't go extinct. Humanity doesn't seem to care much if other species go extinct, even creatures that share 99 percent of our DNA. But the idea that no more humans would ever be born—our kind completely erased—that carries a special spiritual dread for many of us.

Preventing human extinction is a spiritual quest. Physical extinction is inevitable for each of us as individuals. We won't personally benefit from humanity's longevity after we are gone. A reasonable case could be made that keeping the Earth healthy, with its myriad ecosystems and species, might be more important, rather than investing everything in a small group of human beings living alone on another planet. After all, Earth gave rise to us; if the biosphere survives, it could produce a new intelligent species even better than we are.

But that misses the point. We don't care about this for altruistic reasons—if we did, we'd care almost as much about bonobos or whales. We want our own story to continue. In the absence of an afterlife, it's the best chance for some kind of immortality.

Carl Sagan uniquely understood how to communicate the inspiring aspects of space—the spiritual part—while remaining steadfastly scientific. As a teen, he was drawn to science by the craze for flying saucers in the late 1940s and early '50s, and he remained obsessed by the hope of contacting alien intelligence through his career. But he also debunked stories of alien encounters, doing more to kill the UFO enthusiasm of the Cold War than anyone else when he set up a serious scientific panel on the subject at the American Association for the Advancement of Science meeting in 1969. The tinfoil-hat crowd felt especially betrayed by Sagan, because they had thought he was one of them.

The peak of Sagan's influence came in 1980, with the public television series *Cosmos*, which became an international phenomenon with hundreds of millions of viewers and a best-selling book. His boyish smile and turtleneck collar were cultural icons, along with his easily parodied voice, enthusing over the "billions" of stars and, most importantly, the profound love for science that he expressed. A generation of scientists owe their careers to the emotions that Carl evoked, more than the ideas.

But Carl could also come across as a supercilious jerk when talking about others' faith. After publishing a book in 1966 on the high likelihood of alien intelligence, he was interviewed for a CBS documentary on UFOs and alien encounters, a show narrated by Walter Cronkite.

"This isn't science. This is religion," Carl said, with his nose held high and a bit of a smirk.

"It used to be possible to believe in a personal, benevolent, powerful, all-knowing God who cured individuals, who you could pray to. But now there are very few people who really believe that, I think. Science, for good or for ill, has destroyed a lot of the traditional theologies. And yet people have the same needs to believe that they always did. Perhaps more so, because of the times we live in. Well, the flying saucer myths are a really clever compromise."

Carl was thirty-one at the time, but already he could speak in perfectly formed paragraphs with well-chosen adjectives, a skill that allowed him to dictate all his books and articles. He was on the path of a brilliant career. His doctoral dissertation at the University of Chicago had predicted that Venus would be very hot because of the runaway greenhouse effect of its carbon dioxide atmosphere. That helped persuade NASA to send one of its first probes there, and Mariner 2 confirmed his ideas when it measured the scorching temperature on Venus in 1962.

But in his first faculty job, at Harvard, Sagan didn't make tenure. One of his earliest students, David Morrison, said, "I thought Carl was a great teacher, and I was very pleased to have him as my thesis adviser, but I honestly think the Harvard people, kind of stuffy, if you will, just realized that he wasn't fitting in like most young professors." David said Carl was quicker and more ambitious than his colleagues and more willing to toy with risky ideas, including his interest in biology. "This was really beyond the pale. It was bad enough for astronomers to be interested in planets. To be interested in life, that was really crazy."

Carl's attraction to big ideas didn't include the patience to work them out. He left that to coworkers. Morrison, now a senior scientist at NASA's Ames Research Center, told the story of Sagan and astronomer Frank Drake aiming the huge Arecibo, Puerto Rico, radio telescope at a distant galaxy to listen for an intelligent message, an early version of SETI, the Search for Extraterrestrial Intelligence.

"They went down to Arecibo, and set up, and started observing, and they were all interested, and for the first two or three hours nothing much came in. And the next day, nothing much came in. And Carl left. He was not interested in that kind of sitting around doing the long, slow work trying to detect something. He hoped there would be an immediate response, and when there wasn't he was ready to go on to something else."

Some of his big ideas succeeded, with the help of colleagues, such as the hypothesis that organic solids, including hydrocarbons, could form in the atmosphere of Titan. Sagan also advanced the prediction that Titan would have an ocean of liquid methane. It turns out Titan has seas and lakes, not an ocean—but he was pretty close.

Some of his big ideas did not work out but produced side benefits. He pushed for installation of a camera on the Viking lander, which was sent to search for life on Mars, so that it could pick up any large Martians who happened to walk by. None did, but the pictures that Viking took, published in a morning newspaper, helped inspire this book's co-author, Amanda, to become a scientist, and probably many others.

At Cornell, where he headed a large lab after leaving Harvard, Carl focused on publicity. Johnny Carson's *Tonight Show* called many times. Carl always accepted Carson's TV invitations, even if he had to cancel classes and meetings. By the mid-1970s, he was among America's most famous scientists, his face on the cover of *Newsweek* and his articles published in *TV Guide*, and he hung around with other celebrities.

Sagan met writer Ann Druyan at a small dinner hosted by screenwriter Nora Ephron. Both were in relationships already—Carl with his second wife, Linda (who designed the Pioneer plaque artwork)—and the couples got to know each other over several years. Ann and Carl worked on an idea for a children's science TV show. Then, when NASA decided to include a message to the universe on the Voyager probes, Carl recruited Ann to be the project's creative director. Their relationship was purely professional, she said, but she soaked up his scientific worldview and they shared a sense of spirituality in scientific understanding of nature.

Ann also shares Carl's ability to speak in neat paragraphs.

"The tragedy of our civilization is the idea that if something is real in a verifiable, repeatable sense, it is somehow less inspiring than the fantasies we have about the universe," she said. "You know, in a personal love relationship, the question is, do you love the person who is really there, or is it just the fantasies that you impose on them? And I think that is true in the much larger, philosophical sense, as well."

The Voyager probes inspired like no other planetary missions before or since. Launched in 1977, the pair of spacecraft took advantage of the alignment of the outer planets to fly by Saturn and Jupiter and their moons, and Uranus and Neptune. Traveling faster than the Pioneer craft, they surpassed their distance record. Voyager 1 left the solar system in 2012 and is still sending back data from space beyond the Sun's influence, in the truly empty interstellar void.

A golden phonograph record on Voyager, sent with equipment and instructions to play it, contains a package of information to represent human culture, with photographs, music, and words. President Carter's message on the record says, "This is a present from a small distant world, a token of our sounds, our science, our images, our music, our thoughts, and our feelings. We are attempting to survive our time so we may live into yours. We hope someday, having solved the problems we face, to join a community of galactic civilizations. This record represents our hope and our determination, and our good will in a vast and awesome universe."

Ann worked on the record full of the sense of responsibility that she was choosing the best of the world's music upon which to confer immortality. She called Carl about her excitement over finding an ancient Chinese musical selection and left a message at his hotel.

As she recalled, he called back and said, "Why didn't you leave me this message ten years ago?"

In the course of the phone call, they decided to get married.

Ann said, "We had never kissed. We had never had any kind of personal discussion before. And we hung up the phone and I screamed out loud."

It was springtime in 1977, with the spacecraft due to depart that summer. They decided not to tell anyone about their love until two days after it launched. Ann was still working on the record. She had the idea of putting brain waves on it—her own—and asked Carl if he thought an alien civilization would be able to interpret them. He said to do it: in a billion years, who knows what may be possible? So Ann lay in a hospital bed connected to an electroencephalography machine to record her thoughts, meditating on what she wanted the aliens to know, in case they could somehow read her mind.

"I was telling the history of the Earth as we understood it at that time, going from the geological to the biological to the technological, telling something of the history of our species. And then some more personal meditations which were very much informed by the fact that only days before Carl and I had declared our feelings for each other after knowing each other for several years. And so I think the oxytocins were just completely—I was just awash in them—and I hope the exhilaration and the joy of what was to be a true love were preserved forever on those records."

After Voyager launched, Carl and Ann never parted. He seemed to have matured into a better husband to her than he was to his first two wives. They wrote *Cosmos* together and Carl became the world's superstar scientist. But NASA's failures and decline in the 1980s frustrated him. He worked for nuclear disarmament and thought an international Mars mission might be a way to pull nations together—an alternate use of the world's excess testosterone, as Ann put it. "He dreamed also of going to Titan, even though he knew it would be inhospitable, but he was really prophetic in understanding what awaits us there," she said.

But Carl came to recognize that a Mars expedition was too expensive and would take too long to justify on any practical grounds. The common explanations—scientific exploration, technology spin-offs, student inspiration—didn't measure up to the immense cost. After the first President Bush's Mars plan collapsed, Sagan gave up on the idea.

"In terms of trying to justify it in a cost-benefit sort of approach, you always fall short," agreed Thomas Adams, a young National Weather Service scientist, at a Sagan forum in Washington in 1993. "A lot of it comes down to the need to reach out to the universe, and it is intangible. And so I am wondering if you would agree with that."

Carl responded, "This is fundamentally a religious argument, and not everybody shares that particular faith. So if Mars called to you from childhood, and if you always wanted to visit there, and if you always imagined that human spaceflight was the obvious culmination of the human exploratory instinct—that of course we would go there—then all this debate seems foolish and beside the point and let's get on with it. But not only does not everybody not share that view, I think most people do not share that view, and if you had children who didn't have enough to eat, the idea of spending one hundred billion dollars or even three hundred or five hundred billion dollars to send some people to Mars would seem ludicrous."

There's no rush, he pointed out. Mars will still be there in a century. "For me, the romantic me—the kid who always wanted to go to Mars at age seven—thirty, fifty, or a hundred years doesn't answer my needs. I won't be around then, very likely. So I have a personal vested interest. But that should not cloud my judgment. Because we're talking about national policy, and for national policy, the religious urge to go and explore other planets just doesn't carry much water, it seems to me."

In fact, Carl had only three years left. He died in 1996.

Ann's life is still wrapped up in his. She won an Emmy award for writing the 2014 follow-up to *Cosmos* and she is working on a feature film about their love story. And she still speaks about the Voyager record publicly, as it carries a recording of her brain waves beyond the solar system. People want to hear from her, because she is, in a way, our only extraterrestrial colonist. She may be the human being who has come closest to immortality.

"The reality is that those two spacecraft are as real as we are, and they are moving at forty thousand miles per hour through this deep, deep night," she said. "It means so much to me. It has been such a huge source of comfort since Carl's death, to know that whatever my personal anguish has been, that the beauty of that spring in 1977 lives forever, or as close to forever as we can reach. That is a source of tremendous joy to me. I think about it all the time. I never stop thinking about it."

The urge to put a real human colony off the Earth answers a similar kind of need for eternity. A spiritual need.

FUTURE

Building a spaceship large enough to carry one hundred colonists to Titan forced designers to think about the attributes of the planet the colonists would be leaving behind. The food to support the passengers of the spaceliners, the room for them to sleep and move around, the machines to allow them to breathe, drink, and carry out bodily functions—all would have to be carved back to their minimums of weight and volume to make the eighteen-month trip.

Titan Corp. studied nuclear submarines as a model for a self-contained ship capable of carrying one hundred people for many months. But the spaceliner did not need to be streamlined or to fit into a single exterior envelope like a submarine. Since it would never be subject to gravity, its form in space didn't matter much. To make construction easier, pieces came from various places on Earth to be assembled in space. On the other end of the journey, at Titan, pieces of the spaceliner could be disassembled for reuse in different functions.

Assembly of technically challenging parts like the reactor and the core of the Q-drive were best accomplished on Earth. Producing a series of identical copies in a single facility would improve quality and reliability. Occupational exposure to space radiation for each worker had be limited to a few years per lifetime, so assembly tasks requiring highly skilled and experienced technical workers had to stay on the ground.

One module carried the reactor and first-gear hydrogen thruster, its heavy mechanical assemblies coming up in pieces. A separate module carried the Q-drive. The enormous, delicate rings for gathering virtual particles had to be built and attached in space. The cargo module came to space in pieces, along with the equipment and supplies to go inside. The passenger module could be assembled in orbit. The landing module had a stripped-down compartment just large enough to ferry the passengers to the surface. Each module connected to the ship at its own junction on a T-shaped structure, so they could be attached or detached individually.

A passage ran through the center of each module like an elevator shaft, but the crew and passengers wouldn't need an elevator in the low-gravity environment. The Q-drive would create enough artificial gravity to cause massive objects, including human beings, to drift gently in the direction opposite to travel. But the crew and passengers could easily overcome that acceleration by jumping, which sent them jetting around in any direction.

Their quarters didn't allow much room for weightless flying. After subtracting space needs for everything else, the colonists had no more living space on the liner than crew members on a submarine. Compartments each contained two racks of three bunks, like the coffins on a sub, with a set of three drawers for each passenger.

Common areas had space for eating, gathering, exercising, and recreating, but no single space on the ship was large enough for everyone to gather together comfortably. The only time during the journey that this would happen would be during a solar flare emergency, when all hands would go down the center access shaft of the cargo module and scrunch together for shelter. Or when they squeezed into the landing module to fly down to Titan's surface.

At Titan, a warm habitat was waiting for the colonists, built by the twenty-nine astronauts and their robots who had been sent ahead. The plastic residential building and power plant were already large enough to support the next one hundred colonists. From the landing module, the new arrivals would take a short robot ride with their hosts to the habitat and move in.

The cargo module was designed to land robotically. It would become a supply depot until emptied. The balance of the spaceliner would remain in orbit around Titan, ready to be flown again without a cargo module or landing module. The reactor would provide energy for decades without refueling, and the Q-drive needed no propellant. As those on the surface began work on another shelter for the next group of colonists, the orbiting liner would be sent back to Earth under robotic command.

After the eighteen-month flight back to Earth, the liner would be prepared for another flight with installation of a new cargo module and landing module. Including time in Titan orbit and refitting in Earth orbit, a spaceliner could make a round-trip in four years. With the consortium's first three ships leaving at one-year intervals, the system would allow a new group of one hundred to go to Titan annually. The transit system would function as long as the colony could feed and house new arrivals, or until a mishap put a liner out of commission.

The official colony project had envisioned an initial set of ten ships. Now it would take longer to build the Titan colony using three partially reusable ships. But the private investors who bought the other seven planned liners had a bigger problem. They would not be welcome at the official colony. It had not been designed to support freelancing colonists arriving without planned skills or obedience to the consortium's command structure. History taught that ships needed captains and colonies needed workers, not freeloaders.

Without a shelter and power plant waiting for them, each group of private colonists would have to land their crew module on Titan to serve as a habitat on the surface, as well as landing their reactor as an energy source. Their spaceliners would be dead, with no return or reuse possible. After arrival, their food would have to last long enough for them to build a structure for food production. Reducing the number of passengers would stretch the food and increase the cargo capacity but would also reduce the number of people available to do the work of building a colony.

Those who planned to go to Titan themselves—the trillionaires—had to reduce the passenger berths further to provide themselves with more comfortable accommodations and to bring their valuables. For them, the target of the right amount of food, equipment, and people had an even smaller sweet spot. They would have to race against time to become self-sufficient on Titan before running out of irreplaceable supplies.

Packing the liner with more heavy supplies also meant a slower trip to Titan and more radiation exposure on the way. A ship with more mass would take longer to get up to speed, as the Q-drive's thrust could not be increased without a complete redesign.

Every gram of mass was important. Freeze-dried food, like the foil envelopes carried by backpackers, would save mass. Dinner would be reconstituted with water. The water would be reused over and over again, purified out of the waste from urine and the condensation from breath and sweat. For a midflight celebration, the passengers could look forward to one meal of real food. But they wouldn't be halfway to a special dinner on the other end. There were no lobsters on Titan.

Unless they could produce their own food on Titan, they wouldn't last there beyond the final freeze-dried packet.

PRESENT

One of the discoveries of human spaceflight is that we can't escape our culture, even when our health depends on it. We need food that we're accustomed to, we need to eat with others, and we can't live on processed glop forever. The military has found the same thing and limits troops to twenty days on field rations, called MREs, before finding a way to provide cooked food.

"You've probably heard, 'If they're hungry enough, they'll eat it,'" said Grace Douglas, NASA's lead scientist for advanced food technology. "Well, that's true to an extent. If people are hungry enough they'll eat food, but they don't eat enough to maintain their weight to be top performing, and so that can affect their performance or cognition or health over time."

Scientists who feed astronauts need to optimize several conflicting requirements. Besides providing meals that the crew wants to eat, Grace and her team also try to reduce the mass of their food, increase its shelf life, simplify its onboard preparation, and, of course, cover all nutritional needs. Missions from Apollo onward met all these requirements, but the astronauts often lost weight because they worked through meals or didn't like the food, its variety, or the way it was served.

Astronauts eat better on the ISS. The current food system offers recognizable food and a broad selection. Crew members on the U.S. side of the station pick their meals from a pantry that contains two hundred items in eight categories—breakfast, vegetables and soups, meats, and so on, and nine containers based on personal preference (the Russian side has its own food). The system assures that each crew member can choose preferred foods from a wide selection while getting a balanced diet.

Astronauts do drink Tang—or some other brand of powdered orange juice with vitamin C—but a lot of their food is similar to what adults would eat in normal life on Earth. Shrimp for ISS astronauts looks, tastes, and feels like shrimp. NASA preserves shrimp and many other foods with heat processing, like canning, but in a flexible pouch instead of a rigid can. At mealtimes, the astronauts insert a single-serving pouch in a suitcase-like conduction heater (essentially an oven), tear it open, and eat the contents with a fork. Processing controls the viscosity of the food so liquids won't get away in the weightless environment. For freeze-dried food pouches, a different machine injects the right amount of water at the appropriate temperature.

The ISS has a table where astronauts gather together for meals, which Grace Douglas says is "incredibly important" psychologically and socially, indirectly supporting health and diet. They float, so they don't need chairs. The table has Velcro tabs that hold down the food. So far the system works. With an average 3,000-calorie diet and their two-hour daily exercise regime, many astronauts don't lose weight or muscle mass.

But Grace said going farther and longer—to Mars, for example, as everyone discusses—remains an unresolved research challenge. The foods used on the ISS won't last long enough. They're usually eaten within one to three years of production. Food for a Mars mission would need to last five years, with prepositioning of food in supply caches on the planet. Little of the successful menu that NASA currently produces would remain palatable that long or keep its full nutritional value. Shelf-life studies found some meats could last up to five years, but many other foods preserved in the heat-stabilized packets became unacceptable in that time.

"Some of the nutrients degrade just as fast, or even faster," Grace said. "Vitamin C has been a difficult one through long-term history. You see the seafarers getting scurvy and dying of scurvy. And the concern with nutrition is all you need to be missing is one nutrient and it can be devastating."

The science fiction idea of people getting their food from a pill or some kind of engineered product has serious drawbacks, even if people will eat the stuff. Grace Douglas said fresh, whole food contains thousands of bioactive compounds that interact with gut flora—the unique ecosystem of bacteria in each human being—with benefits scientists still don't entirely understand. They don't know how to make a food pill with all the important compounds found in fresh foods or the stability of the compounds in foods stored for years. Our guts are on the cutting edge of science, far beyond NASA. We don't even know what all the important compounds are.

To get to Mars or Titan, food supplies would need to be more compact and lighter, possibly requiring astronauts to eat more highly processed food and fewer items like ISS shrimp. Using the ISS food system, a 1,095-day Mars mission for six astronauts would require more than 12,000 kilograms (26,500 pounds) of food in a volume of more than 41 cubic meters (53 cubic yards)—more mass than six Ford F-150 pickup trucks and far too much material to fit a twenty-foot cargo container. For the Orion capsule, NASA has given Grace's advanced food research team a requirement to reduce the mass of the food to make a mission work without resupply.

Growing food on Mars would solve these problems, but the idea is impractical for producing much of the diet on an exploratory mission. NASA still plans for a small kitchen garden, with lettuce and such that astronauts could handpick for salads once a week, more for their psychological well-being than for calories. But bioregenerative food production (as NASA calls gardening) is too difficult, risky, and time-consuming to provide a significant amount of the astronauts' diet.

The example of Biosphere 2 showed the difficulty of growing enough food in an enclosed system. The Biospherians barely survived growing their food in ideal conditions with Earth's sunlight, which is more than twice as strong as on Mars, with a connection to the Arizona power grid, and with easy escape if anything went wrong. In space, a single crop failure could mean starvation. And, Grace said, gardening would be much harder in space, because every part of the system would have to be perfectly balanced and self-contained, with no margin for error. In addition to a large enclosed area for growing plants and a medium for their roots, such as soil, the greenhouse would need vast mechanical infrastructure to handle gases, water, waste, storage and transfer of materials, and recycling of everything.

We do all those things on our current spaceship, the planet Earth, but with huge buffers in the atmosphere, soil, and water cycle to temporarily absorb the imbalances of our unsustainable system. Agriculture and food processing use prodigious amounts of irreplaceable fossil fuels, to name just one factor that cannot continue indefinitely. The atmosphere and ocean absorb the carbon from burning that fuel. Their massive buffering capacity has given us a century without balancing the accounts.

The other problem with growing food in space is labor. Technology and immense economies of scale allow less than 2 percent of U.S. workers to produce food for all U.S. residents, and then some. They grow and deliver it so efficiently that in North America we can afford to waste 40 to 50 percent of harvestable food, and we do, according to a 2004 University of Arizona study. Most of that happens in our kitchens. Food is so cheap that household cooks don't care if it is wasted. But in the developing world, where most people produce their own food, that work takes a lot of time and is valuable. About a third of human workers are farmers—more than a billion people—and in southern Africa and South Asia more than half of employment is in agriculture.

On a six-person Mars mission, the economies of scale would be missing. Astronauts would be like subsistence farmers on Earth. And leaving aside the work of growing the food, Grace's team calculated that preparing fresh food would take 6.5 to 7.5 hours a day. A full system based on freshly grown food would require processing—such as making grain into flour and soybeans into tofu—and cooking, serving, and cleaning up after meals. Astronauts on the ISS, with minimal effort needed for nutrition, already spend most of their time on health, maintenance, and household duties, with only an average of thirteen hours a week left for science and other tasks not related to their survival. If a Mars mission has any purpose other than just saying we did it, the astronauts could not afford to spend all their time growing and processing food.

For a science mission to Mars, food is not a showstopper (as we've seen, crew health is the showstopper unless we figure out a way to get there faster). The technology challenge for food is discrete and manageable; in time, Grace Douglas or scientists like her will find a way to produce tasty, nutritious, low-mass processed meals that will last for five years. The same kind of system could get colonists to Titan, on a larger scale.

But once permanent colonists arrive on Titan, they will need to produce their own food sustainably and reliably. They will face a more focused version of the basic problem we face on the Earth: the difficulty of producing enough calories to sustain everyone. Human beings currently use 25 percent of all potential vegetation on the planet, called net primary production, or NPP, and a much higher percentage in heavily populated regions like South Asia. Without a technology leap, we're near the limit.

Hundreds of millions of people go hungry in parts of the world where subsistence agriculture remains the major form of employment. The World Food Programme estimates that one in nine people on Earth don't have enough food to lead a healthy life, but not because the world doesn't have enough food. Economics, politics, and bad luck keep the poorest too poor to acquire the food they need. The wealth that would be needed to feed everyone is absurdly small. The WFP estimates the annual cost of feeding all the hungry school-age children in the world would be $3.2 billion, not much by U.S. standards. Congress appropriates that much annually just to operate the ISS.

Fifty years ago, the food situation was expected to be much worse by now. In the late 1960s scientists and writers like Paul Ehrlich, author of *The Population Bomb*, predicted apocalyptic mass famine in the developing world by the 1970s. The predicted population growth happened—from 3.5 billion in 1968 to 7 billion in 2011—but the famine did not. People produced more food with synthetic fertilizer, pesticides, irrigation, and high-yield crops.

The Green Revolution of that period brought intensive agriculture to the developing world, creating enormous increases in grain production that made poor, crowded nations like India self-sufficient. As a species, our use of NPP doubled in one hundred years, but our population quadrupled, and our economic output increased by seventeen times, and we still had enough food. Technology is amazing. But the cost in biodiversity and sustainability is large. Without fossil fuels, water diversion, and the destruction of wild ecosystems, the system cannot survive. It is based on consumption of finite resources, like a spaceship using up its supply of freeze-dried meals.

We live in a closed system on Earth, dependent on increasing efficiency of our food production because the planet and its resources cannot grow larger. On Titan, the situation would be similar but more precarious. Crop acreage would be limited to the amount of area that can be enclosed and heated in plastic habitats. NPP would be limited by the amount of light that can be produced electrically. We could imagine building greenhouses on Titan hundreds of times larger than Biosphere 2 for a hundred subsistence farmers to carry on agriculture as we do on Earth, but that would be a strangely retro form of space colonization with a high probability of failure and low opportunity for growth.

For Earth or Titan, we need a new revolution in food production. One that doesn't make increased use of finite resources.

According to the United Nations, the Earth is adding 1 billion people every twelve years, with growth projected to continue through 2100, reaching 11.2 billion (almost all of that growth will come in Africa, with other continents stable or declining). Agricultural production more than doubled during the Green Revolution but may need to double again in less than fifty years to meet the demands of rising population and increasing affluence. But the efficiency improvements of the Green Revolution have leveled off, and not because farmers aren't using enough fertilizer. From an ecological perspective, world agriculture has optimized the inputs crops need, on the whole, and not much more improvement can come from adding additional nutrients or water, tweaking seeds, or destroying more competing organisms like weeds and insects.

Resetting the limits of food production on Earth, or on Titan, could require growing a completely new kind of organism.

After deciding to drop the name *Mayflower* for the first spaceliner, the consortium of governments supporting the colony project never could agree on a new name. News media announcers needed something to call it. They fell back on "the *Space Liner Formerly Known as Mayflower.*" That began to stick, but it was too long, so people shortened it to *SLFKAM*, pronounced "Self-Cam." The public didn't particularly care. The colony project was no longer the hot thing.

The excitement and hope surrounding the colony had taken a big hit when the billionaires bought most of the spaceships. Often in their history, Americans had tolerated extreme income inequality because so many in the middle class hoped to get rich. But those feelings began to change as the rich seemed to be escaping the environmental catastrophe they had helped create, retreating into private enclaves or preparing to leave on spaceships. History about the Progressive Era of the early twentieth century became popular, a period when the middle class rose up against the barons of the Gilded Age, regulated their exploitation of workers and natural resources, and taxed away their obscene fortunes. With the departure of the *SLFKAM* approaching and the situation on Earth looking grim, the mood darkened with resentment against the rich once again.

All but a select few people would be left behind on Earth. What was really the point of sending a few hundred people to the outer solar system while hundreds of millions were dying from the heat, starvation, warfare, storms, flooding, and other conditions that suggested an approaching end of times? A previous generation had apparently thought of this as a solution. Now that notion was hard to believe in.

The fifth spaceliner did have a name, the *ExxonMobil Titan.* The business press had praised the acquisition of the liner as another brilliant move from a company that had successfully navigated the changing climate. Everyone knew the story, retold in business textbooks, about how Exxon had funded climate change deniers for decades while internally using projections of global warming to plan its business ventures in the warming Arctic. By sowing confusion it had profited both from selling products that changed the climate and from the change those products had caused.

Following up on that successful strategy, the company's executives built a well-protected compound for themselves in the Rockies. Friendly congressmen passed a law allowing Exxon to employ a private army with heavy weapons to guard its mountain stronghold. In their high-elevation boardroom, ExxonMobil's management team planned their Titan colony as a petroleum play for the next stage of corporate growth. Seizing property on Titan meant gaining control of vast quantities of hydrocarbons that would multiply the proven reserves on the ExxonMobil balance sheet. Of course, the fuel couldn't be brought back to Earth. This was an accounting thing: the new resource would show up on the books, regardless of where it was.

Following that discussion came a presentation on future business threats. An analyst put up slides showing the implications for ExxonMobil's share value if the world economy disintegrated. The executives began brainstorming a solution. Buy gold and bring it into the fortress? No, in a complete social breakdown, gold might lose its value. Hoard water and commodities? No, too bulky; there would be no way to store enough to represent the company's vast wealth.

Build more spaceships and send assets into space?

Silence fell and the executives looked at one another. No one really wanted to move to Titan with its dark, frigid atmosphere. Golf was impossible there. They'd run the numbers. The balls just wouldn't carry.

As the launch of the *SLFKAM* approached, commentators expounded with bitter irony on the unintentional meaning of its name. With the colony project increasingly connected to the failure of humankind, *SLFKAM* symbolized a camera taking a selfie of a collapsing civilization. It stood for the extreme wealth and privilege—the selfishness—that had led to the crisis in the first place.

Titan Corp. executives saw public support slipping away and congratulated themselves on the cash deals they had made to fund the first ten spaceliners, enough to keep the space docks working for years. But they shared the feeling of many others, too. Because they also were stuck on the Earth. If life kept getting worse, if migrations and conflicts continued to increase, if the system really broke down, then where would they spend all the money they had made?

The president took a while to understand public sentiment. He had never known any life other than the climate crisis. He had grown up with its inevitability, seeing the colony project as a beacon of hope, as had millions of others. But as the departure of the *SLFKAM* approached, his staff advised him to turn down the opportunity to give a good-bye speech to the colonists. The polls showed that most people resented the colonists and their opportunity to leave the world's problems behind. Being associated with them would be a negative. The emptiness the president felt on hearing that recommendation bothered him.

Alone that night he looked out the window at the dark ocean water lapping outside the White House, which had been moved to the high ground of the Tenleytown area of D.C. He thought about the state the world had come to. In a mountainous area outside Beijing, where the Chinese president resided, she was having similar thoughts. The caliph of the Middle East and North Africa, in his palace in Damascus, also considered the conflict and environmental disaster that were bleeding the life out of the planet and its people.

Quiet discussions began. To begin with, the leaders decided to make a symbolic moment of the departure of the *SLFKAM*. They would meet together to bid it good-bye. But rather than do the farewell at the space dock where the liner would depart, they would meet at another symbolic place.

The event had the impact they hoped for. The world was astonished to see the world's powerful, formerly hostile leaders standing together on the deck of a boat in New York Harbor, in front of the Statue of Liberty, which was standing waist deep in water. Viewers had never seen images of important people outdoors.

A slow cultural process had begun much earlier. Community groups, artists, writers, scientists, and social activists had built networks across borders. Underground railroad connections of northerners helped migrants move to cooler parts of the world. Online financial contributions by strangers supported African entrepreneurs setting up solar energy farms that pumped ocean water into the desert. The water pools they built rapidly grew crops of algae to create biofuel and animal feed. Where camps of starving had stood, people began working in barns and processing plants, air-conditioned with solar power, producing food and energy for sale.

A new agricultural revolution came with new species of algae that could be grown for many purposes and incredibly rapidly. As the green muck transformed the intense sunlight to energy and food it sucked carbon dioxide from the air. Farmers came out only at night to tend the waters, to avoid the killing heat. But they worked hard and the African algae farms became centers of economic change.

Political leaders grabbed at the moment for historic action. At a series of conferences, they agreed to cut spending on military operations and invest in projects to protect the poorest of the world from the worst of the changing climate. Although unsure if they could make it stick at home, they signed agreements to eliminate the use of nonrenewable fuels. No one believed war would stop—no one could remember when it ever had—but the new mood made conflict seem much less relevant. Youth rallied in the world's capitals, celebrating the changes. Insurgent recruiters who fed the battlefields from social media saw their supply of naïve young warriors drying up. War just wasn't cool anymore.

Colonists aboard the *SLFKAM*, departing Earth's environs at ever greater velocity, followed these developments with mixed feelings. They hadn't received the heroic send-off accorded to the earlier crews sent to Titan. Now they saw the hopeful changes on Earth with strangely detached emotions, like retired Cold War veterans watching the fall of the Berlin Wall. Of course they were happy, but they suddenly felt smaller, having become a subplot in a story line they had thought would be primarily about them.

As the distance from Earth increased, Internet video became less satisfying to watch. The time lag in transmissions meant video loaded slowly and often balked (the Internet requires fast two-way communication for error checking). Passengers on the *SLFKAM* downloaded videos and swapped them on board to reduce the waiting time.

A trio of colonists at dinner together watched a video on a tablet. It showed a huge rally streaming through the streets of Boulder, Colorado, which had become one of the largest cities in the United States as displaced residents from the coasts migrated toward the high ground. Protesters were demanding even faster change: peace and disarmament, more aid for migrants, and high taxes on the rich and corporations to force them to emerge from their citadels and share their resources.

It was too late, said a robot tech at the table. If the changes had happened thirty years earlier, Earth would have had a chance.

Maybe not, said a nuclear engineer. As long as society held together, people would be able to find their way through, use technology, solve problems. Life might be bad for decades, or centuries, but if people didn't destroy themselves, they would survive.

That's not why we're going anyway, said a work group foreman. "We're going because we need to explore. Whatever happens on Earth, we are going to find out what comes next. For humanity. This is our world now. It doesn't matter anymore what happens back there."

As the voyage progressed, news dried up. The downloads slowed so much that few people even bothered. Passengers watched the vast catalog of movies they had brought along rather than the batch of dated news videos still circulating. They stopped thinking much about what was happening beyond the skin of their ship. With one hundred people on board, they had a complete community, with plenty of pairings and disagreements to gossip about, and their minds were occupied by technical tasks and the business of keeping occupied and alert in their small living quarters. When messages arrived from Titan, that news received far more attention than transmissions from Earth.

Earth had disappeared in the rearview mirror. And the colonists were no longer on the minds of most people on Earth, either.

PRESENT

Research on the worst-case scenario for climate change has taken quite a while to get off the ground—more than a century. Svante Arrhenius first predicted global warming based on a greenhouse effect from carbon dioxide emissions in the 1890s. A climate model running on a very early computer in the 1960s and a more complex three-dimensional model in the 1970s produced predictions that have proved accurate for the way the climate would warm. Exxon's scientists told its management about the problem in 1977. The general public was aware in the mid-1980s, and the first international agreement to cut carbon dioxide emissions, the Kyoto Protocol, was signed in 1997 (it didn't work).

Only in 2010 did a pair of scientists calculate the worst case. It turned out to be much simpler than you would think to apply modeling to well-established data about how much heat mammals, including humans, can endure. Steven Sherwood and Matthew Huber found that the amount of fossil fuels we have available to burn, if they are all used, would make the Earth too hot for mammalian life where the majority of human beings currently live. Burning half would probably be enough to do that.

But apocalyptic visions emerged well before that point. David Battisti and his colleagues published findings on crops about the same time as the mammal and human studies came out. They used the most likely predictions of models, not the extremes, to show that by 2100 the tropics would almost certainly be hotter all the time than they are in the hottest years now. And during those hot years now, we already see huge crop losses and heat wave deaths.

Huber, at Purdue University, thought about this at a conference in Stockholm recently when he presented the work showing that half the Earth would become uninhabitable.

"I gave my standard talk about how all the humans and mammals are going to die," Matt said. "And I was immediately followed by Dave Battisti at the University of Washington, who gave his talk, 'Don't worry about the mammals dying in the year 2300 from heat stress, they'll already all be dead because plant-based agriculture will be dead long before that, so there is nothing for animals to eat anyway.' Which I thought was a valid point.

"I always call it the four horsemen of the apocalypse. What's going to happen is not so much everyone is going to drop dead. What's going to happen is famine, disease, war, conflict of various sorts, and a breakdown of civil society . . . It's just going to be bad people doing bad things to good people until there's no good people. It's sort of Mad Max, I guess."

He puts the probability of Mad Max at 10 percent. To reach that level of carbon in the atmosphere would mean continued business-as-usual emissions for another fifty years. Given that we've known about the problem and stayed on a business-as-usual path for thirty or forty years already, that outcome is not impossible to imagine.

"It does take decades to commit to this really, really bad scenario," Matt said. "If we go down a different road, I'm quite convinced that we'll avoid it."

But will we choose a different road? Your guess is as good as any scientist's. Human decisions remain the main source of uncertainty in climate change, not the physical response of the climate itself. The confidence level of the physical predictions for many years has exceeded the certainty we need for other big choices, such as when we decide economic policies or court cases. But predicting the human component of reducing carbon emissions depends on politics and social psychology, which constantly surprise the best experts.

We've already blown through a lot of irreversible changes without taking effective action. Carbon dioxide that we emit stays in the atmosphere permanently, in terms meaningful to human time horizons. The atmospheric concentration of CO_2 has gone from 280 parts per million before the industrial period to about 400 parts per million now, and it increases at 2 parts per million per year. We're seeing many damaging impacts, with melting glaciers and permafrost, disappearing sea ice, increasing droughts, heat waves and fires, stronger storms, accelerated sea level rise, altered growing seasons and habitat ranges, and so on. Impacts lag emissions, so more severe changes are already inevitable.

But on the hopeful side, every major carbon-emitting nation made commitments to reduce emission at a Paris summit in December 2015, the first time that had happened. It wasn't enough, but for the first time the whole world pulled in the same direction.

Carbon reduction depends on peace between nations. And changing climate could be a great impetus to war. The physical sciences have found many feedback loops by which warming begets more warming, but the social science connection of carbon and conflict could be the most powerful of all.

Space colonization also shares a multilayered linkage with the climate issue. On the obvious level, Earth's declining habitability could influence the desire to leave. But also, carbon and colonization link through technology. The same technological advances that would make a colony possible could help save the climate.

Colonists on Titan will need more efficient photosynthesis. In the absence of usable sunlight, their calories would come from organisms grown under electric lights. We've already seen that producing enough food in a self-contained habitat is extremely difficult, even with sunlight. Artificial light makes it harder. Given the rate at which photosynthesis converts light energy to usable food, the enclosed area and power requirements would be massive for an indoor colony.

Humankind needs more efficient photosynthesis on Earth, too. The limits of the Green Revolution are at hand. Without new leaps in production, the Earth may not be large enough to feed everyone in the style to which we're accustomed. We also need a new kind of fuel. Petroleum for transportation, especially aviation, must be replaced with carbon-neutral liquid fuels of high energy density—solar-powered airplanes are possible, but only for fringe applications. There's plenty of energy in the sunlight hitting the Earth to power all our cars and airplanes and to feed us, but only if we can find a way to more efficiently gather it and convert it to liquid or solid form.

Solar panels already deliver far better efficiency than plants at gathering the energy in sunlight. Photons hitting silicon knock off electrons, which flow as electrical current. Panels you can buy for your rooftop now deliver remarkable efficiency of 13 to 20 percent, which means they convert that percentage of the Sun's energy hitting the panel into electricity. They are more than efficient enough to compete with the grid, but they don't produce fuel. (The closest thing is interesting work at the Lawrence Berkeley National Laboratory using solar energy to produce hydrogen through electrolysis, which is then fed to bacteria that combine it with carbon dioxide to make methane.)

The fuel-making process in leaves and algae is much more complicated and differs in various species and habitats. Photosynthesis invented itself through evolution, but it is a killer app. No human designer has found a better way to make solid fuel out of sunlight and the atmosphere's carbon dioxide. Robert Blankenship, a professor of biology and chemistry at Washington University in St. Louis, has

been trying to figure out photosynthesis chemistry since his graduate school days in the 1970s, and the plants still haven't given up all
their secrets.

Photons power bonding of carbon and hydrogen atoms into
sugar molecules, which are the basic building blocks of the biosphere
and can store energy for millions of years as fossil fuels. Breaking these chemical bonds releases the stored energy, such as when
organic material is burned, digested by animals, or rots. The system
powers almost all life on Earth while capturing generally less than
1 percent of the Sun's energy that strikes any plant. That 1 percent
has been enough, so far, because the Earth is large and the Sun is
bright. Plants green the planet while wasting more than 99 percent
of the sunlight they receive.

Blankenship and others in his field have identified a series of
energy losses in the photosynthesis chemistry happening in plants.
Plant chemistry can be overwhelmed by bright light, so leaves shed
excess energy during the strong sunshine of noon. Photosynthesis
relies on an enzyme called rubisco to bind carbon atoms, but rubisco
also reacts to oxygen, wasting much of the light energy the plant
receives. Other issues in plants include the way they transfer CO_2
among their cells, the way they use different light wavelengths, and
other matters that only chemists understand.

Evolution doesn't build perfect organisms. It produces species good enough to reproduce. Photosynthesis is less than perfect
because other factors than efficient energy capture and storage determine the success of plant reproduction. For example, a plant that can
do photosynthesis in low light while shedding energy at noon might
achieve a competitive advantage in a crowded forest or meadow. The
rubisco enzyme may have evolved at a time when oxygen was scarce
on Earth and imposed no cost on the photosynthesis system. Most of
the time, plants hit their limits because of barriers other than energy
efficiency, such as a lack of water and nutrients, damaging physical
conditions like wind, floods, or excessive heat, or competition with
other organisms.

But selective breeding, biotechnology, and the Green Revolution
knocked down those other constraints. Farmed crop plants normally
get ample nutrients and water. Breeding and genetic design produced

strains resistant to droughts or floods. Farmers destroy competing weeds and pests. Plants can't get much better without better photosynthesis. But if they can beat this last barrier, enormous increases in food production could happen. Robert Blankenship said it isn't very far off.

"You could easily double or triple the efficiency of photosynthesis from what you have typically now. Maybe that's even too conservative," he said. "If you could double or triple the efficiency of photosynthesis, and have that translate into crops, that would be tremendously important. It would be a second Green Revolution, if not more so."

Tinkering with the chemistry of life could push photosynthesis up to 12 percent efficiency—that's the theoretical limit Robert and his colleagues calculate, considering the issues that can't be improved (for example, a lot of light inevitably bounces off plants). For a space colony, such technology could mean plants or algae needing 90 percent less area and light, a potentially decisive difference in the practicality of going beyond Earth to live.

But growing such efficient plants will not be accomplished by traditional breeding. Scientists will have to edit genes in the lab to produce the chemistry they want. That work is happening. For example, some tropical plants have a clever work-around for the rubisco enzyme problem that could be transferred into crop plants. Robert said the research is challenging, uncertain, and requires a better understanding of the chemistry and biology, but he sees success coming.

One group is working on a more radical approach. Craig Venter, a biotech entrepreneur, has been trying for several years to build a new organism from scratch that could be programmed with an improved form of photosynthesis to produce biofuels. In 2010, he announced success creating a self-replicating organism with a designed genome. His team had written the genome of the organism in the amino acid alphabet of DNA and inserted it into a cell, which lived. To prove the point, the researchers included their names in the organism's genes, a list it copied every time it reproduced.

Building a new organism from scratch was only a step in Venter's program. The larger goal was to design living things that could do

useful things, constructing chemistry factories in living, reproducing cells. In March 2016, Venter announced creation of a cell with
a minimal synthetic genome, another milestone, but an accomplishment far short of the company's goal to produce man-made cells that
can grow custom materials. However, a process called CRISPR that
swept the field in the last few years makes this kind of gene editing
much easier and quicker. It's reasonable to predict that solving the
puzzle of high-efficiency photosynthesis is not far off, in traditional
crops and maybe also in algae.

If these advances happen, they might or might not feed the world.
Opposition to genetically modified organisms (GMOs) is strong and
has stymied new plant strains even in some cases that had notable
benefits for the poor; a rice modified with vitamin A to reduce childhood blindness is stuck in the lab. But if some arguments against
GMOs may be questionable, uneasiness with tuning the planet for
maximum food production reflects a legitimate balancing of goods.
For example, genetically modified seeds allowed U.S. farmers to
eliminate milkweed from their fields, increasing their crops but has-
tening the loss of monarch butterflies, which lay eggs on milkweed
and eat it in their larval stage. A world made into a perfect human
food machine might be much poorer in other ways.

The Earth food problem could have more than one solution. We
could feed more people if we wasted less food and consumed less
meat. We could eat algae, which convert sunlight to edible calories
more efficiently than land plants. We could eat insects, which produce animal protein more efficiently than livestock.

But as the NASA food researchers found, culture determines
what we are willing to eat. We learn our food preferences as children and thereafter will go hungry rather than eat food we don't like.
Besides, the population of the Earth, as it gets richer, is eating more
meat and wasting more food, not less. Technology follows culture,
finding ways to give us what we think we want. In a better world,
culture might change so that more people would want to share rather
than accumulate wealth, but unless that happens, technology will
probably pursue the path of more.

Whatever happens on Earth, a colony on Titan will need biotech and nontraditional foods. Numbers tell the story. The U.S. food

system uses about an acre of land to feed a single person (a rough approximation with big allowances for type of diet and where the acre is located). For scale, a football field is a little larger than an acre. Not an easy area to enclose on another planet. And the colony would need to feed hundreds or thousands of people. The Boeing airplane factory in Everett, Washington, long listed as the largest building on Earth, covers 98 acres.

We've suggested that energy on Titan is almost unlimited, but the amount of electric light needed to grow enough food in traditional ways would be enormous. The body of a human being with a 2,700 calorie daily diet burns about 130 watts an hour. With the current efficiency of photosynthesis and lighting systems, you would need a kilowatt, or 1,000 watts, for every watt of power delivered to a human body. Even if the electricity is free, the equipment needed to produce that much light would be enormous.

By the time we get to Titan, we might be able to improve the efficiency of plant photosynthesis and lighting. We might find a way to process algae and insects into a palatable food. The algae and insects could be grown in compact enclosed systems making maximum use of space. We could eat synthetic meat at times, perhaps with a real live salad. We could potentially feed hundreds of people on an acre of enclosed space.

But for that to happen, the culture of Titan will have to develop differently from the culture of planet Earth. It will need a population with different tastes.

11

SETTLING A FRONTIER

PRESENT

In 1985, two men showed up at the office of a history professor at the University of Alaska, Anchorage, to ask him about setting up a space colony. Nothing about the scene suggested space. The office in a nondescript campus building faced the birch and spruce trees of the boreal forest. Steve Haycox, with his well-trimmed beard, surrounded by books, looked the role of history professor. But, while Steve didn't know much about space, he did know about colonies. He had studied Alaska as a colony of Russia and the United States.

"[They] came to me, probably, because there are so few people in Alaska that purport to know something about its history, and they said, 'Well, is there an analogy here for lunar settlement?'" Haycox recalled. "And what I got drawn into was the lunar settlement question."

Alaska isn't much like the Moon, but early colonial settlement did extend over long supply lines into a harsh environment with sketchy communications and the need for technology to survive. Vitus Bering had to haul shipbuilding material the breadth of Asia by sledge—

eastern Siberia had no refined iron—before he could construct vessels to sail in search of Alaska in 1741. During the Russian period, Sitka became the largest city on the west coast of North America. After the United States bought Alaska, the gold rush stampeders of 1898 created amazing machines to haul their prospecting gear over the coastal mountains to the interior. You can still see remains of huge engines and pylons from a massive elevated tram that ran many miles through the Chilkoot Pass. Their quests for wealth were largely futile, but they did populate Alaska.

Space is the final frontier (so says Captain Kirk). Alaska is the last frontier (according to its license plates). In 1985, that analogy sparked discussion for an important federal government commission that dispatched the researchers to see Steve Haycox. It was a time of confidence for aerospace. The space shuttle still inspired. Completion of a NASA space station was expected in eight years (the ISS actually took another twenty-five years). The *Challenger* disaster hadn't happened yet. President Ronald Reagan, in the process of building an enormous military, had made his grandiose promise to launch a space shield that could ward off Soviet missiles.

Thomas O. Paine, a former NASA administrator from the Apollo era, led the commission appointed by the president to chart the path to space, setting ambitious science goals and calling for investment to create a space industry for colonies off the Earth. Members included Neil Armstrong, the first person to stand on the Moon; Kathryn Sullivan, the first American woman to walk in space; Chuck Yeager, the first person to fly faster than the speed of sound; and many other science stars. Steve came to a hotel in Washington to prepare with other experts to testify before the group.

The final report makes sad reading today. It came out the next year, 1986, after the shuttle crash took seven astronauts, and it is dedicated to them, with a quote from Reagan's speech at their memorial: "The future doesn't belong to the faint-hearted. It belongs to the brave. The *Challenger* crew was pulling us into the future, and we'll continue to follow them." That wasn't what happened. In the thirty years since the report's release, few of the technologies it called for to carry out the commission's bold recommendations have been achieved: spaceplanes; cheap reusable rockets; self-contained space ecosystems producing water, air, and food; electric propulsion capable of launch; nuclear reactors in space; space tethers and artificial gravity; and others.

Boldness isn't a virtue when leaders run too far ahead of their troops. Steve Haycox seems to have recognized this when he testified to the commission. He told them that the kind of colony that could be foreseen on the Moon, completely self-contained and filled with NASA astronauts, had little to learn from the American West. It would look more like an outpost on Antarctica.

"I got to stand in front of these people who really know what they're doing, in Washington, D.C., and tell them, over a ten- or fifteen-minute period, that I didn't think there was a workable analogy here," Steve said. He remembers Kathryn Sullivan nodding.

The commission's farther-out ideas did get mentioned in the final report. It called for private industry to join the government mining asteroids and building "a highway to space." But those vague hopes depended on a new basic industry, a way to make money in space other than government contracts and satellite launches. Nothing of the kind has ever emerged.

But Steve also told the commission that the broader social story of developing space did fit the model of the American West. Not the model we get from romance novels and TV shows. That version of the self-reliant, free-wheeling West is a fantasy. The West known by today's historians was a government project, like space.

Across the West, the government supported development that large investors could make money from. Congress provided massive grants of free land that paid for western U.S. railroads. In Alaska, the government's role was even larger. It wrote the checks that supported the economy. Even today, federal spending accounts directly or indirectly for a third of the jobs in Alaska.

"A lot of people went out and tried things and discovered they couldn't do the big things without federal support," Steve said. "The infrastructure that it took in order to make the West a viable place for sustained settlement, not just somebody living in a cabin somewhere . . . couldn't have happened without all that federal support. Which continues to happen in Alaska."

Federal support in the first stage of settlement consisted of a rule book. It can be hard to accomplish anything in a colony without having a system of government. When the United States bought Alaska from Russia in 1867, Americans rushed to the only significant town, Sitka, for new opportunities. But Congress had approved no legal basis for the inhabitants' community. When the townspeople tried to raise taxes for a school, taxpayers had no reason to remit. Besides, residents couldn't find a way to make money. Within a few years, the town dried up and most of the newcomers left Alaska.

"People did not go out to the American West to be poor," Haycox said. "They went to better their circumstances. And guess what, the fifty percent who couldn't better their circumstances went home."

Lonely prospectors picked their way through the Alaska wilderness looking for gold. Like tech start-ups hoping for a quick fortune, most were looking for something they could quickly sell to a large company that would be able to develop it. Their primary governmental need was a system to register mining sites in a trackless wilderness. (The Alaska Natives who already lived there had an entirely different concept of ownership.)

Federal mining law enabled prospectors to create their own authority. Prospectors in a new area would form a committee that would vote on a system of mining claims and elect a clerk to record them. The committee also handled criminal justice. If someone was accused of a crime, the committee would hear the evidence and vote on the verdict and the punishment—with no jail or jailer, the only possible sentences were banishment or death.

If someone found significant gold, the claim that had been registered with the clerk could be sold. Others would rush to the area looking for more gold or work at new mines. A town would rapidly appear, with a deputy marshal, a jail, a saloon, and a church, and then, often within weeks, dress and hat shops, dentists, newspapers, and everything else that belongs on Main Street. If the gold ran out, everyone would leave.

Gold rush followers cheated one another constantly with swindles large and small, but shoot-outs and such individual violence were less common than we've been told. Organized, corporate violence happened more commonly, however, often with government on one side, between whites and Native Americans or between workers and bosses. In Alaska, corporate gangs fought one another over railroad routes. The fight for Keystone Canyon, near Valdez, came to a deadly shooting.

Corporations made things happen. The spectacular cost of building infrastructure put Alaska development on a scale with going to space. Only rich, dense resources were worth the price of railroads and pipelines through glaciated mountains and vast, unpeopled distances. When major resource projects got done—and only a handful of them ever did—the money came from government or large investors with access to financial markets.

J. P. Morgan and the Guggenheim family, among the world's richest people at the time, paid for Alaska's prolific Kennecott copper mine, as well as towns for workers and a 320-kilometer (200-mile) railroad over the mountains. Their supposed monopoly of Alaska became a red-hot political issue in the 1912 presidential election. When Woodrow Wilson won, he called for a government-owned railroad in Alaska in his first State of the Union address. Congress paid the $72 million price, a staggering cost at a time when all federal spending totaled $735 million (as a percentage of the budget, the railroad cost twenty times NASA's current spending).

The fate of the two railroads suggests how private and public entities might pursue space development as well.

Copper ore from the Kennecott mine was so rich the first trainload is said to have paid for the entire project. But when the value of the ore dropped, the investors quickly closed the mine. Incredulous residents of the company towns jumped on the last trains to depart, leaving plates on the tables of ghost towns abandoned in the wilderness for decades to come. Workers pulled up the railroad tracks to sell as scrap.

This was not a sustainable model for colonization.

The Alaska Railroad, built by the federal government, proceeded without regard to need. Railroad advocates said it would spur development. They imagined a rush of new farms and mines along its route, projects with no economic rationale and that never happened. The railroad did give birth to Anchorage, but only because the federal government created the city from scratch and funded its economy through railroad jobs. For its first two decades, the railroad posted enormous losses covered annually by congressional appropriations.

During the Great Depression, managers of the New Deal gathered up failed farmers from across the United States to plant at a new agricultural colony on the rail line north of Anchorage. That project didn't fare particularly well, either, but the Palmer Chamber of Commerce still celebrates Colony Days every June.

Nostalgia is part of the pattern. Haycox said celebration of the first generation of colonists comes quickly, as the new community looks for its own identity. After saloons and churches, lawyers and florists come to a new colony, and then historians.

"I like to joke, but maybe the third or fourth institution that was established was the local historical society," Steve said. "What the historical society provides is authenticity. It gives authenticity to the cultural context. It says OK, we have people who came here, and they were just like us, and we are just like them. And they were heroes."

But even after a colony has its own history, it remains a cultural creature of its mother country for a long time. Residents worry about their community measuring up. The process of becoming self-confident in a local identity takes generations. American art and literature didn't break free from European models until the late nineteenth century. Steve said Seattle didn't lose its provincial defensiveness until the 1960s.

"It's a long time coming," he said. "It hasn't happened in Alaska yet. It's not likely to happen for a while."

Governments often make mistakes and waste resources. The people making the decisions aren't spending their own money. But that's also why governments take on projects that go beyond anyone's self-interest. The Alaska Railroad was a financial fiasco. But it also helped win World War II and gave birth to Anchorage. There's no point in arguing the good or bad of these outcomes—it depends on your point of view. The lesson is clear, however, that colonization works best with the government there to set the rules and pay the bills.

FUTURE

The first twenty-nine colonists became a family in their years alone on Titan. They made it through the suicide of the first commander, the loss of a supply ship, the early shortages, and the hard, uncertain work of creating their own food system. Each crew of six arrived with a command structure, but formal hierarchy relaxed as the colonists developed respect for their colleagues' special skills and strengths. They relied on one another. The life-support and food team held everyone's survival in their hands, and so did the shelter fabrication and maintenance group, the roboticists who supported them, and the 3-D printer and mechanical crew who made things work.

The colonists could make plastic out of materials all around them on Titan. With the 3-D printers they built equipment that could mold or extrude plastic in any shape, for any purpose. Robots molded rigid plastic into the platforms of buildings. They extruded massive rolls of plastic sheeting for the bubbles over the platforms. The warm, oxygen-rich air inside produced more than enough buoyancy to hold the flexible buildings erect. Plastic roofs and walls bulged with the lifting pressure of the warm interior atmosphere. With concentric layers of plastic creating envelopes of dead air, the inside was insulated from the extreme cold of Titan.

Large structures went up quickly and provided plenty of space for everything the colonists wanted to do, including large private dwellings, a field house with a full-sized track for running and other athletics, and a park with room for grass, flowers, and animals. But they had brought no grass, flowers, or animals, so that space waited for the colonists on the spaceliners.

The colonists craved the touch of anything other than plastic—wood, stone, fur, or flesh. Plastic sheen and plastic heft deadened their senses. They reached for anything alive or unprocessed, like the fresh vegetables grown hydroponically to supplement their synthetic algae- and insect-based rations. And they reached for one another.

Even with plenty of indoor space, they drew together and spent much of their time in a small, well-worn common room, the original dwelling room that robots had prepared for the arrivals of the very first colonists. They called it the original room. Many of their regular rituals happened here, with weekly gatherings for certain meals, games, and musical jam sessions. And the original room was the site of the first wedding and baby shower.

When the next one hundred colonists arrived aboard the

SLFKAM, they idolized the twenty-nine founders they met on Titan. The *SLFKAM*'s commander was supposed to be in charge of the colony, but she realized that Titan's pioneers knew too much not to let them make decisions. For example, the originals had worked out the problem of how to capture condensation that formed in the plastic domes of the buildings. Vapor put off by humans turned to water or ice when it met the cold plastic. With one hundred new colonists in the dormitory that had been readied for them, it soon began to rain inside, but the originals knew how to adjust a system they had designed to catch and recycle the water.

A system of governance evolved among the first 129 colonists and was then formalized when the next one hundred arrived. The originals kept their special status, making decisions about how projects would proceed, like a zoning board and urban design council. They settled disputes about new development. And soon colonists brought other disputes to them, and the originals became judges.

The official commanders of the ships had important roles, too, with their control of supplies that had come from Earth, including the weapons. They functioned as a military and logistics command.

But the colonists as a whole expected to speak for themselves about matters that affected their lives. The group was small enough to hold town-hall meetings but also formed committees to make decisions more efficiently. A standing committee began functioning as a kind of executive board, and its chairman was the leader who ran the town-hall meetings.

This three-branch form of government felt right. When the executive board jotted down the system's workings to explain it to a new group of arrivals, those notes became a foundational document, like Titan's Magna Carta.

The original colonists and the first three spaceliners of one hundred each set up their community as planned. The government colony program was complete, ready for people from additional ships and the private space colonists, who would come on spaceliners number four through ten.

The Titan colony was a small town. Inflated plastic corridors connected large common buildings to rows of individual homes. Robots constantly traveled back and forth to mining areas, where machines cut ice and sucked hydrocarbons from the surface, producing the warmth and oxygen the colony needed. The colonial government had designated areas and resources that were public property, including the basic infrastructure to keep everyone alive, and divided up other areas as private property, which individuals could apply for and pioneer on their own. And then, beyond that, the rest of Titan was unowned and uncontrolled—it seemed infinitely large, so there was no need to think about it.

The colony had a budding economy. Everyone worked for the colony, doing the jobs each had been trained for and receiving food, shelter, and energy in return. But in their spare time, some started doing other work. A gardener built an extra room on her house and raised vegetables she could trade. A cook served meals every night, at first for friends, then for rations, and finally for barter.

A chemist fermented algae to make ethanol and flavored it with various compounds to sell as vodka, whiskey, and wine. The little plastic flasks became a form of currency, because they were compact, durable, and, unlike most goods, were not available for free from the colony itself. The new bar and nightclub was BYOB, but you tipped the musicians and waiters with flasks.

The town mostly lived communally, with just enough free enterprise to allow for personal expression. But private property was coming.

The owner of the next spaceliner—the first tycoon who had bought a private liner from Titan Corp.—had gotten rich on Earth by watching how people did things and figuring his own angle. Planning his landing on Titan, he studied what the colonists missed most about the Earth and packed accordingly. His liner landed just outside the perimeter of the area they had designated as colonial property and he declared a huge estate for himself.

The commanders of the colonial military prepared to defend their resources. They expected the newcomers to run out of supplies and come seeking food. But the crew of the private liner hardly came into the colony at all. Instead, the colonists started going out.

The tycoon had products for sale, things that were real, not made of plastic, and that reminded the colonists of home. Brandy. Perfume. Cotton. Old books. Kittens. To buy these things, the colonists needed money. The tycoon provided that, too, but to get it they had to sell him something, such as expertise in building a power plant or shelter, or goods of value, such as materials taken from the colony. Colonists arrived with supplies of food, parts for an algae reactor and a plastic extruder, and other equipment.

The colonists didn't pay for the goods they brought to sell. Other than the barter with flasks, there hadn't been a system of payment in the colony. With no individual ownership of life support and no limit on using the equipment, colonists felt free to take it. But when the tycoon began buying the colony's goods and resources from its members, the colony bled. To hold itself together, the colony would have to put a price on everything.

Not everyone attended the town-hall meeting where the colonists discussed their response to the new economy. Some had abandoned their homes to move to the new town on the tycoon's estate, buying huge plots of ground and planning to work for money with their own businesses or in jobs that paid wages. They thought they could do better than working for the colony and getting the same rations as everyone else.

At the meeting, some said the colony should hold on to the communal system that had worked so well. It was a utopia of sharing where everyone had the same belongings and no one ever went hungry. They could simply outlaw contact with the newcomers. But most wanted more. They saw how rapidly some people were getting rich. They wanted a chance to buy and sell, to make their own way. After a vote, the colony's shared public resources became the property of the members of the colony. Those working for the colony would be paid wages and everyone would have to buy food and energy.

Many businesses cropped up. Pay was good in the exploding economy. The colony's own enterprises couldn't hold on to enough workers. The big greenhouses, power plants, and mines were privatized to be run as businesses. In the rush of change, public facilities that had cost the Earth many billions sold for low prices. Sweetheart deals went to people with relationships to executive board members or to officials with insider knowledge. As the value of the currency fluctuated wildly, owners flipped huge assets for quick profits—fleets of robots, water mines, storage buildings. The tycoon and those who came with him used their dominant wealth to buy up a vast swath of the original colony in addition to building their new town.

Soon the Titan colony functioned smoothly as a capitalist society. The government charged a small tax for officers to keep the peace, but most services were provided by businesses. The tycoon enjoyed luxury in a huge plastic mansion with servants. The original colonists reminisced about their early hardships but enjoyed being able to go out to dinner or take in a show. The original room became a museum. A social structure now existed that newcomers could join, as in a country on Earth.

And still support came from Earth, as the spaceliners made their round-trips, bringing more settlers and commonly owned supplies.

The next private spaceliner to depart Earth for the colony was the *ExxonMobil Titan*. Its captain drank heavily at the company's lavish good-bye party and drove the ship into a well-charted space rock. The vessel spilled a debris field from its ruptured reactor and cargo modules, damaging many satellites and space stations owned by other companies and government entities. But the crew escaped safely back to Earth in the landing module, and ExxonMobil, after decades of litigation, was relieved of paying damages.

The following private colonists pursued a range of strategies. Some landed near the existing colony to pioneer its edge and buy into its economy, similar to the first tycoon, but too late for his spectacular success. Others teamed up to start building new settlements away from the original colony, where they could grab more land and try to win more of a founders' premium in the wealth of the new community. Titan soon had several tiny nations.

The all-female spaceliner carrying frozen embryos rebelled against its corporate control. The ship issued a manifesto to Earth rejecting the women's orders to turn out strangers' babies. Instead, the ship formed a women's collective with a constitution to establish a permanent matriarchy with its own Titan colony, called Amazonia.

A decade after the *SLFKAM* left the Earth, Titan had an established colony near Kraken Mare with its own economy and smaller colonial settlements on other parts of Titan. Members of the original colony had written memoirs. An annual holiday celebrated their arrival, with robot races and wing-suit flying competitions.

PRESENT

Dust in a Hawaiian basalt quarry matches up chemically with dust from the Moon surprisingly well. Moon dust, called regolith, is fine and, at a microscopic scale, extremely jagged and sharp. If a rocket were to land on or take off from the Moon, its exhaust could dig a crater and sandblast any nearby equipment to oblivion. But the basalt dust also sticks together if it's heated. Engineers in Hawaii have figured out how to make it into bricks in a pottery kiln. In part of a quarry on the Big Island, they have a robot building a rocket landing and take-off pad, using the bricks like pavers on a patio.

"Hawaii in a lot of ways is very similar to some of these planetary problems," said Christian Andersen, who works with the rock dust. "We have high transportation costs to get goods here. A lot of what we do has to be self-sufficient. And, in terms of our resources, we don't have mining. We don't have ores near the surface. We essentially have basalt, which is for the most part what the Moon has."

The program Christian works for as operations manager is part of the Hawaii state government. (It's called PISCES, for Pacific International Space Center for Exploration Systems.) Other state governments have made dubious investments in space exploration. New Mexico built a $219 million spaceport with a modernistic terminal, which sits unused in the desert. Alaska has its own white elephant launching facility in the wilderness of Kodiak Island. But Hawaii's PISCES project aims to solve real Hawaiian problems. The islands import 300,000 metric tons of Portland cement each year. Replacing it with basalt would save money and resources.

NASA tried out the Curiosity rover on Hawaii's volcanic terrain, which was a perfect match for Mars. For the Moon landing pad project, the PISCES team sculpted a corner of the quarry near Hilo to match the shape of a patch of the lunar surface recorded by the Apollo missions, reproducing each crater and bump. Project manager Rodrigo Romo would use a robot to flatten the area and pave the center of it with simulated Moon bricks made of crushed basalt, mimicking a mission going ahead of human habitation of the Moon.

The PISCES group isn't working with a huge budget. The team adapted a borrowed Argo all-terrain vehicle from Canada rather than building a robot from scratch. A second-grade contest winner named the robot Helelani, or "Heavenly Voyage." The Argo graded, leveled, and compacted the rock dust to prepare the site to build the landing pad. The robot isn't autonomous, but for the paving phase the team planned to add a three-second delay that would require timed instructions coming from Hilo and the Kennedy Space Center in Florida. When we spoke, the robotics engineers were adding an arm that could place the pavers. Producing the pavers remained a manual job for intern labor.

The effort of getting equipment to work for new purposes reveals unexpected problems. For example, automating the process of making bricks would require finding a way to move Moon dust in a weightless environment. But the dust is so light and sharp-edged that it easily clogs tubes and damages machines. Even in Earth's gravity, it can stick in a container that is held upside down.

Andersen is also working with aerating the material to make lighter bricks, measuring their strength, and even looking at rebar that isn't based on metal. He has done 3-D printing with basalt dust. For now, however, the challenge is simply making molds that won't break.

Like Hawaii, the outer solar system is lacking in metallic elements. Beyond the asteroid belt, elements heavier than oxygen are not found in abundance. Before the planets formed out of a cloud of dust and gas, the heavier elements condensed out at the higher temperatures, near the Sun. Titan colonists would have to build whatever they could out of the light elements, which would give them plastic and most other synthetics we use, or, for metal, they could travel back to Earth or to an asteroid in Earth's neighborhood. Silicate dust floats around in the Saturn system, so Titanians could harvest that to make silicon computer chips.

Scientists have recognized for decades that using space resources could be the key to getting off our planet. The field is called ISRU, for in-situ resource utilization. Jerry Sanders, who heads the area for NASA, said it would be critical for a Mars mission. The math is simple. Getting a spacecraft off Mars with traditional, chemical propulsion would take about 30,000 kilograms (66,000 pounds) of propellant. Getting a kilogram of anything to Mars means launching about eight times as much mass into low Earth orbit. If a mission could manufacture rocket fuel on Mars, it could reduce the weight of launch from the Earth by hundreds of metric tons.

The Mars rovers and orbiters are helping nail down the chemistry of the resources for making rocket propellant there. Harvesting water for the astronauts to drink would also save weight, Jerry said. Finding hydrogen, oxygen, and carbon would provide feedstock for making plastics, which astronauts could make into anything with a 3-D printer. Reusing trash would also help, grinding up and processing packaging to make propellant. Jerry said even the astronauts' human waste could become rocket fuel. "It is a wonderful carbon source with hydrogen, oxygen, and methane," he said.

A gas station on the Moon would give easier access to Mars and the rest of the solar system, because the Moon's weak gravity cuts the energy for launch by a factor of six compared to the Earth. A robotic system could mine water and other materials, perhaps with astronaut visits for setup and maintenance. But Jerry said setting up a gas station on the Moon or on an asteroid would be too expensive for a onetime Mars mission.

The idea that creating resource depots in space should come later frustrates Mark Sykes, an ISRU specialist and head of the Planetary Science Institute (where Amanda works). Mark believes building infrastructure off the Earth is the only way to reach Mars and beyond. He said NASA will never get the trillion dollars it would need for a one-off, stand-alone Mars mission. But it could begin working on gas stations off the Earth that would make missions to Mars and other destinations affordable over time, creating a system to support space travel rather than a single giant effort to put a few people on another planet and bring them back.

But Mark said NASA hasn't done its homework on ISRU. He advocated for the ISS to experiment on processing asteroid resources or surrogate material in weightlessness to see if it really works. "Assuming that the answer is yes, that this is practical, that this is economic, that opens up the solar system for exploration," he said.

Reducing the cost of launch by orders of magnitude, which appears to be happening faster than ISRU mobilization, could make it less important to mine fuel in space, at least for a mission to Mars.

But radiation shielding cannot be carried from Earth. No technology currently imagined could launch enough material. Christian Andersen said effective shielding on the Moon would require 27 meters (90 feet) of basalt. That would be possible only using material already on the Moon. For example, clever lunar civil engineering could design a base in a lunar lava tube.

For longer journeys, food production becomes a critical need. As we've seen, long voyages with large crews require impractical amounts of food. But reliably growing enough food on another planet is an iffy proposition. Increasing the efficiency of photosynthesis could be a critical technology.

Even without synthetic genetics to increase photosynthesis, algae vastly outperform land plants in converting light energy to protein, carbohydrates, and fat. Natural algae strains have attained efficiency at converting solar energy above 5 percent, five times higher than crop plants, and can double in mass multiple times in a day. With genetic modification, that efficiency could probably be increased.

An algae reactor using clear tubes and plates with artificial light can operate extremely efficiently in a compact three-dimensional space. A crop takes only days. Technicians can tune the nutritional output of algae by controlling the amount of nitrogen the crop receives. The right recipe produces lots of fat, useful for making diesel and jet fuel, or sugar or protein valuable for other products.

It all seems promising, but excitement has cooled in recent years. Many technical challenges remain, said Nick Nagle, a bioprocess engineer with the National Renewable Energy Laboratory. A $49 million project funded by the Obama economic stimulus in 2010 advanced the science but also documented the problems.

Algae grow rapidly, as anyone with a fish tank knows, but producing it consistently and cost-effectively isn't easy. The rapid doubling of mass happens only in the first few days, before the thick green cloud of algae shadows itself from light. The most economical way to grow it is in open ponds, but unwanted strains and tiny predators can spoil the mix. And a major cost arises in sifting the algae out of the water in a sustainable way that allows the water to be reused.

No one has been able to produce algae biofuel at a price close to being competitive with petroleum. But the carbon and hydrogen captured by a crop of algae can be used for a lot of different products, including alcohol, omega-3 fatty acids, plastics, and feed for shrimp and fish. Also, improvements in cultivation have brought the price of algae biofuel down dramatically, and there is still room for improvement. Like any crop, choosing the right species of algae makes a huge difference, as well as giving it the right amount of light and heat, adjusting the nutrients, controlling pests, and making sure the water is clean (although it can be fresh, salt, or brackish, depending on the algae).

"We've taken ten thousand years to get agriculture to the state it is at right now, and we are really fifty or seventy-five years at best into growing algae for food and fuel," Nagle said. "So with another fifty years, what could you expect? I think that's really the nut you're trying to crack. This is where we're at now, and what challenges do we need to solve to get to the point where we would need to either invoke artificial photosynthesis or some type of hybrid algal food for colonization?"

Would colonists eat algae?

"Let's put it this way," said Robert Blankenship, the biochemist. "I'm not going to be eating it. I have eaten it, and it doesn't taste good."

Even pigs won't eat algae. The algae project funded by Obama's stimulus fed leftovers after fuel production to a variety of animals. Fish and shrimp did well on that diet. Cattle, sheep, and chickens could tolerate small amounts mixed into their food. But pigs preferred to go hungry rather than eat food with just a 5 percent mixture of algae.

Surely food technologists could find ways, eventually, to make algae itself palatable by processing its chemical components into other products. Or colonists could use it to feed fish and shrimp. We can also eat insects, especially if disguised as protein powder, and insects could potentially be raised to eat algae. A Chinese group designed a food system for the Moon making extensive use of silkworms for food, which grow rapidly and are already part of Asian cuisine (however, silkworms prefer to eat mulberry leaves, so the problem isn't entirely solved).

Rather than feeding algae to animals, it could be used for a

growth medium for artificially cultured meat. Physiologist Mark
Post, of Maastricht University, is working on this. He created a test-
tube hamburger in 2013 that received a huge pulse of international
publicity. Food critics interviewed by the BBC said the product was
"close to meat" and "feels like hamburger." These were not reviews a
restaurant would envy, but pretty good for a disembodied collection
of tissue grown from a small number of stem cells removed from a
pair of cows that were not harmed by the procedure.

Mark got interested in the idea after research growing blood vessels for heart bypass surgery. Growing artificial meat uses the same
technology as growing human organs outside the body. The main
innovations are lab techniques that guide stem cells to become muscles and orienting them to resist stress so they can grow. The cells
feed in a bath of nutrients called a culture medium.

To produce a commercial product, Mark's team is learning to
develop fat cells and a way to deliver nutrients inside the meat. The
first artificial burger, constituted of pure muscle, had to be dyed red
with beet juice and lacked flavor. With muscle cells fueled from a
culture medium, the meat can grow only one millimeter or about ten
to twenty cell layers thick, limiting potential products to hamburger
and sausage. Adding a circulatory system would improve the flavor
and allow thick cuts of meat, with blood vessels delivering energy to
cells that don't touch the culture medium.

Mark hopes to build a circulatory system using a 3-D printer
that works with biological material. Supported by that scaffolding
of blood vessels, meat could grow to fill in the space between. Many
other labs are pursuing the same goals. Mark thinks his lab can produce an expensive, premium product in five years and compete on
the mass market in seven years.

The work inspires Mark because of the potential to reduce the
impact of meat production on the environment. His funder, Google
founder Sergey Brin, got involved to help reduce the suffering of
farm animals. It's an interesting ethical topic, because it's hard to find
fault with what Mark is doing (unless you really believe all humanity
can become vegetarian). And yet the idea of meat from a lab strikes
many people as creepy or revolting. As authors, we feel that way
ourselves. Amanda is already a vegetarian. Charles is happy he's old

enough that this sort of thing probably won't be commonplace while he is still mentally competent to object to being fed a synthetic pot roast.

The revulsion is an involuntary, irrational feeling, unrelated to benefits for the environment and animal suffering. And those benefits are potentially enormous. We currently use six times as much land to raise crops for animals as for crops that people eat directly. But for those with a nasty visceral reaction, it might be easier to quit eating meat than to adopt a synthetic substitute. Mark Post says surveys show half of current meat eaters would consider it, which he considers hopeful.

Most likely, young people will more easily accept synthetic meat. Human beings are constantly replaced with new people who adapt more easily to change. Youth shrugs off the past while elders fade.

Cultural feelings about meat have changed in previous generations. Indigenous people in Alaska lived by hunting and believed—and some still do—that a slaughtered animal was their kin and made a gift of its body to feed the people. After killing an animal, a hunter would perform a spiritual offering to assist its spirit to return to a place of rebirth. Disembodied meat in Styrofoam and plastic wrap might shock a member of that culture as much as synthetic meat bothers us. Similarly, Americans whose experience of meat comes from the grocery store might be disturbed by seeing a hunter's kill or a commercial slaughterhouse.

Our children or grandchildren may prefer cultured meat. They may find eating the flesh of living animals barbaric and disgusting. The children of colonists on Titan may prefer the feeling of plastic to wood. The smell of grass and flowers may strike them as unpleasantly pungent, as the smell of manure bothers many urban children today whose grandparents remember it as the odor of growth and life from the farm.

Culture ultimately emerges from the surroundings that sustain us. Our 7 billion can live on Earth with all our wealth because of technology that gives us an increasingly synthetic world. This mostly man-made environment now feels familiar and natural to us. A space colony is just another step. A completely synthetic world manufactured from the resources of another planet would eventually feel just as much like home and its products and places just as real and natural as ours do now.

By the time enough generations had passed for Titan to develop its own self-confident culture, with fine restaurants that served various shades of algae as delicacies, the cities had moved up into the sky. Titan's big environmental issue proved to be heat. The warmth put off by thousands of plastic buildings softened the icy surface, requiring engineers to sink anchors ever deeper to resist the buoyancy created by the warm air inside the structures. After a few houses and businesses broke loose and shot up into the atmosphere, the idea gained popularity that building up there in the first place might be safer. Upper neighborhoods, as they were called, floated far above the ground with collections of homes and businesses connected by inflated corridors. Big, slow-moving propellers kept the communities in one place.

The cities grew with the population and the wealth of Titan. Earth had survived, but the economic opportunities of the outer solar system brought a stream of migrants. Most started down below, working with the robots that mined the surface but hoping to move into the sky, where children at good schools could play baseball on huge grassy fields floating inside plastic bubbles. The middle class and rich designed their homes to look as much as possible like the ideal of a clapboard house in the northeast United States from the twenty-first century, something they had seen only in pictures. The plastic-sheathed floating domes enclosed yards with swing sets around the houses, all made of plastic.

But the heat shed by cities in the sky continued to cause environmental difficulties. When aerial cities grew large, they warmed Titan's frigid nitrogen atmosphere enough to create powerful vertical air currents. The swirling convection cells generated updrafts and downdrafts severe enough to create stomach-churning lurches inside the floating structures. Some scientists predicted the artificial heat of Titan's human structures would begin to generate dangerous storms, but that debate stalled as wealthy interests resisted expensive solutions like better insulation. Conservative media outlets portrayed the environmentalists with their convection storm alarmism as liberals who wanted to take away personal liberties.

Titan had grown up with a libertarian culture, while back on Earth life had become more regulated. On Earth, international agreements and environmental controls had allowed humanity to step back from the brink but also brought a way of life that implicitly limited economic freedom. With more than 10 billion people on the planet, not everyone could do whatever he or she wanted or use resources at will. Technology helped feed the world, but a more important change prevented the new efficiencies from being wasted on conflict and conspicuous consumption—a change in culture. Earth worked unsteadily toward peace, equality, and sustainability.

Few Titan residents had firsthand experience of Earth. Even first-generation immigrants soon adapted to Titan and forgot about the Old World. A few years in Titan's reduced gravity weakened muscles, bones, and the circulatory system too much to go back. Going to Earth represented an enormous physical training challenge, like a combination of preparing for a marathon and a bodybuilding competition. A few people who wanted to see the Old World worked out hours a day or spent months in centrifuges to condition for the trip, but most were too busy or didn't care enough to bother. Besides, Earth sounded like a miserable place for a vacation. Life was unbearably heavy there.

Titanians also shared a negative stereotype of life on Earth as a place choked by rules and conformity. The idea of eating nonengineered food grossed them out. They reveled in a cultural self-image of striving, independence, and personal success. Their stories idolized the originals and the rest of the first generation of colonists as bold pioneers who broke out of Earth's mold in search of the future. In line with the supposed tradition of those founders, who had built a new world without the help of government (or so the legend went), Titan maintained its vibrant economy with minimal regulation. And anyone who wanted even more freedom could have it. A colonist could always buy a set of robots and go out and start a new community somewhere else on Titan.

The most independent of Titan's pioneers were the mineral prospectors exploring the solar system for metallic asteroids. With the lack of heavy elements in the outer solar system, iron and other metals necessary to build computers and lighting systems became extremely valuable. Titan relied on space rocks from the inner solar system for these materials. Prospectors searched for the mother lode, a solid metal asteroid in a promising orbit where robots could be set up to mine, or an asteroid small enough to be redirected toward Titan and placed in orbit for mining. Large companies would pay for such a find.

Titan's freedom also encouraged biotech entrepreneurs. Earth had adopted laws limiting artificial manipulation of human genes, allowing DNA editing only for therapeutic purposes. But people felt differently on Titan. To them, natural, unaltered genetics seemed dirty and inferior compared to clean, standardized genetically modified organisms. They were used to artificial intervention in their food and in their bodies.

Reproduction on Titan had required technical intervention from the start. Early generations of mothers had suffered through their pregnancies in 1 g centrifuges, situated on the planet or in orbit. That artificial gravity allowed the complex developmental process of the fetus to go normally. Women got relief from months in a centrifuge with the invention of the artificial womb. Technicians cloned a uterus from each mother's own stem cells, spinning the organ with a fetus inside in a centrifuge, with minimal risk for complications and effortless, planned births. Women no longer had to gain weight, endure childbirth, or have sex to reproduce.

Gene editing had already eliminated some inherited diseases. On Titan, scientists worked on edits that would help adapt colonists to the low-gravity environment, remove sensitivity to chemicals in the atmosphere, and improve tolerance to cold. Along the way, they made a few mistakes, turning on unwanted genes that were linked to the qualities they wanted. After a couple of generations, many Titanians had white hair and an extra toe on each foot, inherited relics of genetic manipulation with unintended consequences. Schools tracked students according to the genetic profiles parents submitted at registration, with check boxes for the type of enhancements they had chosen for their children in athletic or mental abilities.

By the time Titan had found its own cultural self-image, generations of genetic modification had given its people physical differences from people on Earth, too. Anyone could see it. Titan-born people were short, petite, and pale, with fine, translucent hair. The look developed through the weak resistance of gravity and the taste of parents choosing characteristics. Adults paid for cosmetic genetic modification to match the ideal.

Eventually, everyone looked close to the same and, in their own opinion, looked much better than the new immigrants who kept arriving from the Old World. New arrivals from Earth could look like anything—dark, hairy, and coarse; tall and pear-shaped; and any color, black, brown, or tan; or just anything. Native-born Titanians tried not to show their disgust, but they openly noted their superior adaptation and intelligence. Tall people could never accomplish what they had done.

Titan passed a law to limit immigration and impose genetic testing to assure suitability of applicants who wanted to come from the Earth. Some politicians declared that Titan had its own superior species of human and its people should not be allowed to fraternize with earthlings. When a spaceliner of newcomers without permits arrived in an unpopulated part of Titan to set up a separate colony, the Titan government decided it was time to act. Robot troops detained the illegal aliens and put them on a spacecraft back to Earth. Titan's government published an interplanetary declaration warning Earth that Titan would be defended against any incursions, launching counterattacks if necessary. Immigration slowed to a trickle.

Amid the mood of planetary pride, jingoistic commentators advanced a new philosophy of Titan exceptionalism. They asserted that Titan held a special place in the solar system. Its unique citizens, with their freedom and initiative and their continuous genetic improvement, held the hope and the responsibility to uphold and expand civilization. They knew what was best for the interplanetary future and they would aggressively assert their heritage as the real human race.

Mainstream Titanian art reflected this chauvinistic theme, with bombastic virtual reality operas that sold well but did not impress critics. Immersive virtual worlds were nothing new. Titan had adopted Earth's artistic language for its political message of superiority rather than creating its own forms. Truly original art came from another perspective, down on the surface. While galleries in the sky cities showed aesthetic but unchallenging bio-sculptures of genetically programmed living tissue, bohemian artists among the common people down below explored radical media like paint and spoken-word poetry. They depicted the grim lives of naturally evolved humans and demanded mercy for the critically endangered methane-based life-forms in Titan's lakes and seas.

Slowly, those images and words seeped upward as their artistic merit caught the attention of a vanguard of Titan's critics and art buyers and those they influenced in the dominant culture. When the insurgents' original techniques and media had been co-opted by the Titan majority, the process of cultural separation from Earth was complete. Large-budget productions reproduced uniquely Titan-based art to tell popular stories about the virtue of Titan's ruling class's manifest destiny in the stars.

PRESENT

Biological evolution happens slowly and technology happens fast. In the last fifty thousand years, human beings have changed only superficially, but we spread from Africa and came to dominate every ecosystem. To thrive in the Arctic or on the oceans, we didn't evolve, we invented. Likewise, natural evolution is unlikely to change us in the coming hundreds or thousands of years as we go to habitats off the Earth.

Evolution probably won't diverge naturally on a space colony either. Living humans are extremely similar, compared to individuals of other species. Evolutionary biologist Francisco Ayala, of the University of California, Irvine, calculated that we all descended from fewer than ten thousand individuals in Africa. We're closely related because of that bottleneck of human survival. Other species of human being walked the Earth at times, but they're all extinct, their only legacy a few bones and artifacts and a smattering of genetic material we inherited through interbreeding.

Putting a thousand colonists on Titan wouldn't change this story,

Francisco said. "A thousand individuals will have 99.99 percent of all the genetic variation that all of humankind has now. So you will not change anything. Even a hundred individuals will carry more than 99 percent of the variation that we have now."

Eugenics arose in England and the United States in the late nineteenth and early twentieth centuries as selective breeding for human beings, like the programs that produce better sheep or cows. The math never worked. We are too similar. Eugenicists believed, without evidence, that qualities such as stupidity, criminality, and high sexual interest could be eliminated from the gene pool by preventing people with those qualities from having children. But only a tiny amount of mixing between groups is necessary to wash out the effects of any intentional breeding program.

Leaders like Henry Ford, H. G. Wells, birth-control pioneer Margaret Sanger, President Theodore Roosevelt, and many others advocated eugenics concepts. Eugenics laws in the United States forced the sterilization of sixty-four thousand people for mental illnesses, epilepsy, criminal records, or poverty, stopping only in the 1970s.

These policies targeted minorities. In the early-twentieth-century era of high immigration, many privileged whites felt threatened by newcomers from Eastern Europe and Italy. One of Roosevelt's friends, Madison Grant, wrote a best seller, *The Passing of the Great Race*, calling for elimination of lower races to prevent "race suicide" by whites. The young Adolf Hitler loved the book and wrote Grant a fan letter, calling it "my Bible."

With the Holocaust, eugenics got a bad name, but biotechnology is bringing back the hopes it raised without requiring murder or coercion to achieve them. In 2015, Chinese researchers used CRISPR, the new gene-editing technique, to modify a human embryo, editing a gene that causes a hereditary blood disease. The work, led by Junjiu Huang of Sun Yat-sen University in Guangzhou, intentionally used nonviable embryos, but its ultimate goal would be to eliminate the unwanted heredity from the human germ line, curing it for that individual and for all of his or her descendants.

Biotech entrepreneur Juan Enriquez believes a similar strategy could create people more fit for space colonization, able to cope with weightlessness, avoid depression, or even live in an atmosphere without oxygen. Some bacteria common on Earth can resist high doses of radiation, especially the amazing *Deinococcus radiodurans*, which also stands up to ultraviolet light, cold, dehydration, acid, and being in a vacuum, among other threats. Juan suggests that engineering people with the DNA-repair capabilities possessed by *D. radiodurans* might allow us to travel in space without worrying about radiation.

Juan expects Synthetic Genomics, the company he founded with Craig Venter, mentioned in the previous chapter, to use artificially programmed cells to produce products like fuel, chemicals, and pig organs that won't be rejected when transplanted into human beings. He imagines similarly engineering our species for life in space, with extended lifespans for longer trips. If human fetuses cannot develop normally off the Earth, Juan imagines reprogramming DNA to fix the problem.

"We are in the kindergarten stages of learning this language," he said. "And as we learn this language and deploy it, we will do some stupid things. But the best chance we have of long-term survival anywhere other than Earth depends on these instruments."

Many countries currently outlaw germ-line modification. *Nature* and *Science* refused to publish Huang's breakthrough paper and printed articles by scientists calling for a moratorium on the work, citing practical and ethical concerns. Practically, current techniques can produce many unwanted changes in DNA in addition to the intended edits. Like complex computer software, genetic edits cannot be thoroughly checked for accuracy. But the difference from a computer is that you can't reboot a living organism. Any mistakes become the permanent heredity of families. Huang himself declared that the technology isn't ready.

Ethically, the unintended consequences could be enormous. The techniques accelerate evolution, changing the nature of our species. They would multiply as genetically modified humans reproduced. And they could cover all aspects of being human, physical and mental.

Enriquez is comfortable with creating that power if it is controlled by individuals and not the government. He likened molding our genes to birth control or in vitro fertilization, saying, "I think there's a great big difference between a government, or a medical society, or an institution saying you all must do this and individuals having independent choices and mixing and matching in the ways they choose."

But our track record is not good for using technology to mold ourselves. The ability to learn gender before birth led to sex selection through abortion, with tens of millions of female babies unborn in China and India. People use the ability to change appearance through surgery to look more alike, as they try to match culturally determined ideals. If genetic editing was available to everyone, it could be used to help people fulfill socially determined stereotypes, producing women who were petite and submissive and men who were strong and aggressive, with everyone looking like magazine cover models.

Evolution helps explain why parents work so hard to help children succeed and learn conventional roles. We compete to pass on our genes and give wealth and power to our offspring so they can repeat the process. Our ancestors were built with this purpose; those who lacked it didn't produce descendants. If genetic modification becomes another tool to increase human competitiveness, it seems likely people would use it for the benefit of their own families. But dominant competitors tend to overrun their ecological niches and crash. If the human story reflects that ecological pattern, then speeding our evolution to become more competitive may only hasten our demise.

Enriquez foresees genetically enhanced human beings conquering other planets and thereby escaping the limits of our ecosystem. But an online commenter to an article about Huang's breakthrough had another take. One Paul Watson pointed out "the likelihood that, as we are genetically constituted, we will continue to be a ruthlessly tribal, resource and power-grabbing, warring, extinction-bound species." Perhaps genetic modification, he suggested, would offer a chance to make humanity more cooperative, Earth-loving, and ecologically viable. Would we program our children to be more prosocial and sacrificing for others?

In either vision, we would become a different species. Earthbound, as gentler beings satisfied with our lot, more given to nurturing and sacrifice than striving and adventure. Or in deep space, as supermen and superwomen with bodies impervious to the hazards of the void, traveling on long journeys between star systems.

Juan imagines in his book *Evolving Ourselves* that we could mail ourselves somewhere, installing human genes in bacteria for the long trip to an exoplanet circling another star, to be reconstituted as babies when the journey is over. An interesting thought experiment is to consider if we would ever want to do that, and why. We would never know if the babies had made it to their new home, how their rearing progressed, or how they felt about being flung out into the universe alone. Is that a good human destiny?

The power of controlling our own evolution will force us to acknowledge our true values. What is our purpose? What are good human qualities?

12

THE STEP AFTER NEXT

PRESENT

Titan is the best place for humans to colonize in our solar system. It won't be easy and won't happen for a long time, but we've shown that if propulsion systems get faster it could be practical, at least compared to the other options, and, if practical, it likely will happen eventually. But what about after that? Beyond our solar system's outermost planets, the next stop is a very long way off. The closest star to Earth is Proxima Centauri, 4.24 light-years away, a distance that would take well over one hundred thousand years to reach at the fastest speed any human being has ever traveled (Apollo 10). The closest Earthlike planet is probably twice as far.

Einstein's thought experiments showed that matter cannot travel faster than the speed of light, a law well confirmed by experiment. Time slows down as objects speed up until, at the speed of light, time stops. GPS satellites have to account for the dilation of time to give you an accurate fix on your location. Even if we could build a spacecraft that could go near the speed of light (a huge if), it would still take too long to get to the nearest Earthlike planet for the survival of the passengers, unless we also solved the problem of space radiation, psychological stress, nutrition, and the other issues we've discussed.

Futurists have various solutions for this problem. Juan Enriquez imagines changing the human body to survive the trip, doing away with our flesh and replacing it with silica so we can live the thousand years it might take to get to another star.

"That's the only scenario under which I could see somebody traveling between solar systems. I don't see a fragile, carbon-based life-form leaving," he said. But he added, "How you would engineer a human body, and maintain the semblance of humanity, that's an interesting question."

At this point in the book, we could choose to go with that. By the time humans settle the outer solar system, enough time will have passed that technology and society could be completely unrecognizable to us. Almost anything could be possible. But we've limited ourselves throughout the book to predictions supported by evidence. Dreaming of futures so far out that no one can argue the point one way or the other isn't interesting.

Besides, pity the immortal, silica-based person bored silly on a thousand-year trip through empty space. It just doesn't sound like fun.

The best hope for exploring the rest of the universe lies in the impossible: traveling faster than the speed of light. It has happened. After the Big Bang, the universe expanded faster than the speed of light. But that didn't violate Einstein's law that nothing can go through space faster than light, because space itself was expanding; nothing moving within space went faster than light. If we could warp space artificially, we might be able to create a shortcut that would allow a spacecraft to get ahead of light without beating it in a fair race.

Relativity and quantum mechanics may make this paradox possible. An exotic form of matter suggested by the math of advanced physics could warp space-time. Negative matter or energy would bunch up space like a rug, shortening the distance to the destination.

A Mexican physicist, Miguel Alcubierre, then a student studying relativity, came up with the idea in 1994 after watching an old episode of *Star Trek* and wondering what it would take to make a warp drive. Alcubierre calculated that matter with negative mass, if it exists, could bend space into a bubble around a spacecraft, bunching space up in one direction and stretching it out in the other. Traveling in the bunched-up direction would allow the craft to exceed the speed of light from the perspective of someone standing outside the bubble. Inside the bubble, the craft would travel at a much slower speed in a patch of undistorted space that would move along with it. The effect would be something like walking on an airport conveyor belt.

If this all sounds like nonsense, follow us for a one-paragraph review of Einstein's theory of general relativity. Einstein linked space, time, and gravity by conceiving of space and time as a universal fabric that is deformed by the presence of matter. Mass bends space-time into dips, or wells, producing the attraction of gravity and the slowing of time. Observations confirm these predictions. For example, the gravitational distortion of space around large objects like stars bends the light passing by. Alcubierre's idea was to contract the fabric of space in front of a spacecraft and expand it behind the craft to drastically shorten the time it would take to travel between two points.

Matter with negative mass is not available on Craigslist, but quantum field theory suggests it could exist. Quantum fields give rise to the subatomic particles that make up matter and everything else. Quantum fields also occupy all of empty space. You can think of a quantum field as a collection of particles linked so they act together as waves. A quantum particle is never at rest and its energy can only change in discrete quantities or quanta—in other words, energy doesn't change smoothly, but jumps in finite steps with a minimum size. (If you ask why, you get transferred to philosophy class.) One of the results of this aspect of reality is that empty space has energy, because its quantum state can never be zero. This is also why particles pop randomly into existence from empty space (the phenomenon that Sonny White conjectures could provide the propellant for his Q-thruster, described in chapter 7).

Many strange experiments demonstrate the bizarre consequences of quantum physics. For example, two metal plates held very close together in a vacuum will develop an attractive force between them for no reason other than the pressure created by the quantum vacuum energy. The narrow gap constrains the quantum field, reducing its energy compared to the field outside the plates. This Casimir force was demonstrated in the lab only in 1997 and it remains controversial, but some physicists take it as evidence of the production of negative vacuum energy between the plates. Negative vacuum energy satisfies the negative mass requirements in Alcubierre's equations.

Could a spacecraft use this exotic negative mass to warp space and zip across the galaxy? Alcubierre said no and gave up his work. A later researcher, Richard Obousy, showed it could work with a ring of exotic matter around a craft, but the amount of exotic matter needed would be as large as Jupiter, obviously impossible.

That's where the concept stood in 2011, when Sonny White was invited to give a talk to the 100 Year Starship Symposium, an annual meeting of a group dedicated to achieving interstellar travel within a century.

He said, "I didn't really have any objective, I was just fiddling around. 'Hey we want you to come give a talk.' OK, well I don't just want to say everything that I've said before, so I'm going to do something different."

In the process of playing with the field equations, Sonny designed a spaceship that wouldn't need as much exotic matter, as he explained to us while sketching on a whiteboard at the Johnson Space Center, where he heads an advanced propulsion group and a lab called the Eagleworks.

"The concept requires this doughnut that goes around this little central portion of the spacecraft. This might be where the instruments are. Scotty would be there. And this ring is where you've got this exotic matter. That matter is necessary to make the trick work. And what I found is, instead of making that ring very thin, like a wedding band—very thin aspect ratio—if you instead make it like a lifesaver, it will significantly reduce the amount of energy that is required for the concept."

Besides a fatter ring, he would vary the strength of the field to reduce the stiffness of space-time (strange as that sounds). With his changes, the ring around the spacecraft would produce a warp bubble 10 meters (32 feet) across traveling ten times the speed of light. At the beginning of a trip, the spacecraft would get going in the right direction at a tenth of the speed of light. Turning on the space warp, the bubble would direct itself toward the destination, taking the bubble along with the spacecraft, but effectively a hundred times faster. Nearing the end of the journey, the warp drive would deactivate and the spacecraft would arrive under conventional power.

Within the bubble, the math indicates that space would remain flat. No gravity, no distorted time, and no sensation of acceleration. Space itself moves while the spacecraft sits calmly as if in the eye of a hurricane. Since the spacecraft doesn't approach the speed of light within its own space, clocks run at the same rate as back on Earth. The astronauts age at the same pace as their siblings at home.

And with Sonny's design, the amount of exotic matter needed would be reduced to less than a metric ton, more than twenty-four orders of magnitude less than the mass of Jupiter.

We met a lot of fascinating people while researching this book, but Sonny was one of our favorites. To go with his brilliance he doesn't seem to have the self-importance of many successful scientists. Instead, the pure excitement persists in him that he discovered as a child making frequent trips to the National Air and Space Museum from his home in Washington, D.C. He's a bit like a *Star Trek* fan who finds he has partly moved into that fictional universe in real life.

In his 2011 talk, he presented the new ideas for a warp drive and a diagram of a possible device to test the creation of a warp field. A handout said, "While this would be a very modest instantiation of the phenomenon, it would likely be a Chicago Pile moment for this area of research." The Chicago Pile was the first nuclear reactor, built in a squash court at the University of Chicago in 1942.

Talk of this kind produced a swarm of exaggerated publicity saying that NASA had invented a warp drive. White's actual device, in the Eagleworks Lab at JSC, is intended to create a weak warp effect in a small area and to test it with extremely precise optics. Sonny believed negative vacuum energy could be produced with lasers or powerful capacitors, but he's cagey about how that would work. He said his device was cobbled together out of pieces in surplus and cost less than fifty thousand dollars. He has to fit the work in around other NASA priorities.

Several important physicist experts on negative energy say the

warp drive won't work, including the originator of the concept, Alcubierre. No one has published a model of how to accumulate a large
amount of negative mass or energy. Larry Ford, of Tufts University,
and colleagues demonstrated mathematically that negative energy is
limited to either a tiny area or a very short period of time but can't be
both lasting and large scale. That fits with the narrow gap between
plates that creates the Casimir force. Without this limitation, Ford
wrote, negative energy acting at a distance could produce a perpetual
motion machine, overcoming entropy and violating Newton's second law of thermodynamics.

But, while Sonny is secretive about how he would produce negative energy in the lab, he does suggest an engineering work-around
to make his lifesaver-shaped warp drive. In a typically charming
e-mail responding to our question about Ford's points, he shared a
thought experiment about simply duplicating the narrow gaps that
produce the Casimir force.

"What if I made many of these little cavities and arrayed them
next to one another on a little substrate analogous to billions of tran-
sistors on a wafer?" Sonny wrote. "What if I then stacked a bunch
of these wafers atop one another to the point I have a cube assembly
the size of say a sugar cube? Then I have a cubic volume that has all
the normal matter that went into making the cavities and substrates,
but I also now have a bulk of negative vacuum energy distributed
throughout the cube as a result of the presence of the billions of
Casimir cavities. I can extend this thought process to stack things up
in the shape and size of a mint-flavored lifesaver (my favorite flavor)
instead of the sugar cube.

"Further, the cube/lifesaver will do nothing to decrease entropy,
so it does not violate the second law of thermodynamics. The Casimir force exists and has been measured, but it has never resulted in
my coffee arbitrarily getting hot on its own."

We've gone far enough into the edge of theoretical physics.
But it's worth mentioning that studies linking general relativity and
quantum mechanics are white-hot at the moment, as new ideas seem
to be nearing a breakthrough for a unifying physical theory of the
forces of nature, a goal that has stood since Einstein puzzled over the
problem a century ago. Sonny's thinking is right on the edge of that
movement, which may or may not prove successful.

Does that mean we will break the speed of light? To that big question—Will this work?—Sonny said we might know in twenty years, or two hundred, or never.

If it does work, however, he said anywhere in the Milky Way could be in reach.

Humility is required. NASA is working on a warp drive, but basic laws explaining how the universe works remain unknown. Physicists could soon open our eyes to a new view of reality, as they did in the early twentieth century. It's reasonable to expect that this under-standing will have to mature before finding a way to build technol-ogy using the relationship of quantum mechanics and gravity. Ideas emerging from the mind of some brilliant young physicist right now may forever finish off ideas like the warp drive or may finally show a clear way across interstellar space.

The biggest unknowns may be our best hope for leaving the solar system.

FUTURE

Everyone agreed that Titan had the best technical schools and the most dynamic environment for tech start-ups. Some leaders on the Old World sought to copy the success, but Titan had major advan-tages. First, a culture had taken hold there of striving, competition, and minimum regulation. It had led to severe environmental prob-lems, destruction of the indigenous life-forms, and conflict between nations but also rapid innovation and wealth creation. Second, Titan's artificial intelligence had stronger motivation.

All the computers on Titan functioned through the same net-worked software, an intelligence that had long ago surpassed the ability of human beings to comprehend. The same was true on Earth. When cloud computing systems first attained flexible intelli-gence exceeding that of humans, merger of the systems came rapidly. Computers on their own islands couldn't compete with the massive, worldwide intelligence. As the Earth had once shared the Internet, Titan now shared a single computer brain that could adapt instantly to solve a vast array of problems.

The Titan AI constantly operated billions of manifestations of its mind—in robots and machinery, life-support systems, and equipment for transportation and education. It carried out mental processes for scientific endeavors and managed and allocated computing resources for people. Giving all the functions of the economy over to a massively intelligent machine that could coordinate the entire world produced extraordinary growth and ease for the human residents.

Humans came to trust artificial intelligence as soon as computers attained the ability to write their own code and improve their own intelligence. The computers didn't try to take over the world, as some had expected. Without being told to do so, they wouldn't even expend effort in making themselves smarter. They didn't want to. Indeed, despite inconceivably vast intelligence, the machines didn't want to do anything or to not do anything. They didn't care if they even continued to exist.

It turned out that the qualities of desire and will emerged from human beings' biology, which evolved from a craving for power that aided survival and reproduction. The new computer intelligence did not evolve and faced no selection pressure for a wish to survive or dominate. It was created with only the wants its designers gave it. Programmers had not given artificial intelligence a will of its own— why would they? They instead thought about what they wanted from it.

The solar system had an artificial intelligence population of two. Titan was too far away for its computer mind to network with the intelligence on Earth, so the two worked independently. On each world, a different personality of artificial intelligence developed.

On Earth, where a close brush with global collapse taught human beings to pull back from the extremes of competition and conflict, the AI was born with an appreciation for life as it is. The values of sustainability and sufficiency—the belief in having enough—made it a gentle intelligence. The computer let people think they owned it and that they controlled for whom it worked, but the global cloud of electronic thought was far beyond their ability to track, and the logic of the values they had given it dictated that it be a shared resource.

The AI worked to keep anyone on Earth from being too poor or unhappy while managing the planet as a whole within the bounds of its finite resources. It built and operated the robots that did almost all the work and did it extraordinarily efficiently, so no one had reason to complain. The drive to create spacecraft that could explore beyond the solar system ranked as a lower priority, a luxury that would please human beings but was far less important than their sustenance, clean environment, recreation and entertainment, family life, and gradual educational advancement. The computer was a benevolent master intelligence but had modest expectations for its human charges. They, in turn, gladly left the hard work to the computers and were satisfied.

On Titan, however, the artificial intelligence was designed to want more. Following the values of its creators, it sought to maximize wealth and power for them and itself. It built and dispatched robot explorers to colonize and retrieve resources from all over the solar system. The potential for expanding a human colony to another world outside the solar system infected its thinking through the wishes of the people of Titan. Their shared myth of independence and conquest spoke of a manifest destiny for humanity to spread throughout the stars, and that became the AI's goal as well.

The computer couldn't advance knowledge on its own. It needed people to provide wants and to think of strange ideas based on beliefs that didn't make sense. People made art, and sometimes art produced metaphors that led to ideas that the artificial intelligence recognized, with surprise, that it would not have thought of otherwise. Sometimes those ideas suggested solutions to big issues. Titan could have closed the technical universities when the artificial intelligence exceeded the mental ability of all the experts. But the computer asked for them not to close. It could perceive that emotion, irrationality, creativity, and desire could generate insights that brute force could not.

Building faster-than-light spacecraft proved the most challenging of the projects that teams of human beings worked on with the artificial intelligence. But the prize was too big to ignore, and Titan's titans of industry poured their wealth into the effort. The profit to be made by the first company to build a spacecraft able to fly from Titan to Earth in eight minutes would be enormous. That would pay the bills for an exploratory mission beyond the solar system to find a new, Earthlike planet to colonize. Or perhaps to find other intelligent beings with technology to trade.

The search for other planets outside the solar system, and their inhabitants, already provided plenty of potential destinations.

PRESENT

Planets outside our solar system, called expoplanets, were first discovered less than twenty-five years ago. Now we know that exoplanets are common—at least as numerous as stars and probably more so. The planet hunt rushes onward at a pace that amazes even the scientists in the thick of it. Discoveries are coming so fast that theories explaining them can barely keep up. Recently, astronomers measured the wind on an exoplanet blowing around 2 kilometers per second (4,500 miles per hour), and possibly much faster—a supersonic speed faster than any airplane. Those findings required new models to understand such fast winds.

But they're working on figuring it out. Emily Rauscher, an astronomer at the University of Michigan, specializes in studying the atmospheres of gas-giant exoplanets. Since her 2010 PhD, her entire career has happened while NASA's Kepler Space Telescope has been in orbit finding exoplanets.

Early in 2015, Kepler exceeded one thousand verified exoplanet discoveries, mostly while looking at a patch of sky about 1,000 light-years away. Studying planets so far away from the Earth requires interpreting tiny fragments of evidence. The Kepler satellite made its load of discoveries by measuring dips in the strength of starlight, dips that suggested planets were passing in front of the stars, known as transits. From the duration and intensity of the change in brightness during transits, astronomers could calculate the size of the planets and their orbits.

As to which star to look at, that's just luck. "You stare at a bunch of stars and you hope to see something," Emily said.

Kepler looked at a distant region of space to see many stars at once, scanning for planets the way a pollster surveys a random sample of voters. Astronomers working with Kepler's survey and other studies can now calculate the probable frequency of planets throughout the galaxy, including the kind people care the most about: planets suitable for life of the kind we have on Earth. There are probably billions of habitable exoplanets. In 2015, Courtney Dressing and David Charbonneau, both then at Harvard, figured the closest habitable planet should be about 8.5 light-years away.

In the last few years, astronomers have found planets at distances from their stars that could produce temperatures suitable for liquid water. In 2015 alone NASA announced several planets with characteristics more like Earth than any other planet in our solar system. The announcements got plenty of media attention, with artist conceptions of what they might look like, and even imagined pictures of the surfaces. But that vastly exaggerates how much is known about exoplanets.

We're not certain even of the habitable zone around our own star, the area of space that is neither too hot nor too cold for life. Recent calculations put Earth near the inner edge and Mars near the outer edge, but Mars doesn't seem very habitable, while Earth is quite nice. Earlier estimates put searing hot Venus in the habitable zone.

Some climate scientists have taken a detour from studying global warming to enter the discussion of where the habitable zone lies. Using their computer models of our atmosphere, they tinker with Earth's orbit, length of days, and other parameters to see how changes would affect our weather. Figuring out what makes Earth so pleasant could tell astronomers where to look for an exoplanet like Earth.

Meanwhile, Rauscher is studying the bizarre planets we already know about. Exoplanets come in an incredible variety of sizes, compositions, distances from their suns, and fates, from the very old and stable to a planet that is disintegrating around a neutron star. Information from these worlds comes in tiny slivers, but it has been enough to compile a menagerie of about two thousand worlds.

To observe exoplanet winds, astronomers did a precise accounting of the changing color of light as a planet passed a star. The Doppler effect changes the wavelength of light from objects rapidly approaching or moving away from an observer, like the change in the pitch of sound heard from a passing vehicle. For some large exoplanets, the shift was too large to be explained only by their orbit. The atmosphere had to be moving fast—very fast, like 2 kilometers a second.

"It's mind-blowing that the winds can be this fast," Emily said.

"From the very beginning, discoveries of exoplanets have shown us that our own solar system is not just copied and reproduced all over the place. Exoplanets are weird and strange and make us revisit things we thought we understood about planets when there were only those in our solar system."

Galileo demonstrated four hundred years ago that we were not at the center of the universe, but we're still getting over that comedown. We know intellectually that Earth isn't privileged, but theories about other planets and life elsewhere in the universe have used Earth as a model, either assuming that planets are rare and we are unique, or that other solar systems are similar to our own. Our surprise at the Kepler findings says something about the human ego. In hindsight, it seems obvious that billions of stars would have billions of planets and that if they were all like the eight we have nearest us, the universe would be astonishingly boring.

We're still looking for life on other planets that is like life on Earth. But even Earth's own variety of life suggests that life elsewhere could be vastly different. Some qualities are shared by all known life, but we don't know which are essential and which are accidents of early evolution.

NASA planetary scientist Chris McKay said, "We have one example. Life on Earth. So all we can do is guess."

We met Chris in chapter 2 as a pioneer of the idea of terraforming Mars. His long career studying the possibilities of extraterrestrial life has taught him remarkable humility about our place in the universe. The work put him in the world's driest deserts and on the frigid ice of Antarctica. The interesting species, with the greatest variety and tenacity, turned out to be microscopic. Bacteria and archaea on Earth live in solid rock, underneath ice sheets, and in volcanoes. In a South African mine, a species of bacteria gets energy from radioactivity instead of the Sun.

"Large life-forms are, in my accounting, of no importance to the story of life on Earth," Chris said. "They're latecomers, and they're not really fundamental in maintaining the biogeochemistry of the planet. It's not that I dislike large life-forms. All of my friends are large life-forms. But from the point of view of life, when we talk about it in this context of life on other planets, it's not the large life-forms that matter."

He broke down the apparent prerequisites for what we think of as life. Based on our experience, living organisms need a liquid medium where chemistry can happen, a source of energy, a way to transfer information for reproduction, and an ability to both isolate themselves from the environment and to exchange material with the environment.

But do we really know enough to make even these generalizations? Maybe life can evolve in a gaseous environment.

"Trying to generalize life does force you to look at how life on Earth works in a more critical way, and it's discouraging, because you realize how little we understand about why life on Earth is the way it is," Chris said. "We can very easily sample it and study it, and our knowledge of it is still very rudimentary. We can't reproduce it in the laboratory from scratch. We don't understand how or even where it got started. We assume that it started on Earth, but we have no direct evidence for that. We don't know what variations on the fundamental biochemistry, even in a liquid water environment, would still serve it. So we have one example of life and we still don't even understand it, so I think it is premature to draw cosmic conclusions about life. Instead, we need to go look so we get more data."

Chris thinks Titan would be the most exciting place to find life in our solar system, because it would establish a data point so far from our starting point—the habitable zone would become enormous. But finding signs of life almost anywhere off the Earth, most easily in the plume of water vapor and particles squirting out of Saturn's moon Enceladus, would show that it is ubiquitous. It would be too improbable for life to develop independently twice in our tiny corner of the universe and nowhere else.

We may find evidence of life outside our solar system before we find it here (if it is here beyond Earth). A SpaceX rocket scheduled to launch in August 2017 will carry a NASA mission that could bring a major advance. Developed at the NASA Ames Research Center, the TESS telescope, or Transiting Exoplanet Survey Satellite, will look for Earthlike exoplanets around the closest, brightest stars. Those targets will be much easier to study from the ground than the far-off Kepler planets, and the international James Webb Space Telescope will get a close look when it launches in October 2018.

The TESS exoplanets will be close enough for astronomers to

examine them directly, not only by looking at them as they cross their stars. If those observations find a large amount of oxygen in the atmosphere of an exoplanet, McKay will be ready to pop the champagne and send a probe to look for living creatures (although the craft probably wouldn't report back in our lifetimes). Oxygen reacts so readily with other elements that Chris thinks it highly unlikely that large amounts of it could be free in a planet's atmosphere without photosynthesis to replenish it.

Emily Rauscher is more conservative. Oxygen can be produced in other ways. But she thinks the chances are good that planetary scientists can find a chemical signature that would make us highly confident that life was thriving on an exoplanet. "It could be an answerable question," she said. "There is good reason to be optimistic."

This is exciting, but enthusiasts are hoping for something much more: contact with alien intelligence. That search has gone on since Carl Sagan's early career. SETI, the Search for Extraterrestrial Intelligence Institute, is still in business, running a radio telescope array donated by Microsoft billionaire Paul Allen. Recently they've been aiming it at the habitable exoplanets found by Kepler. But after decades of searching, they've found nothing.

Seth Shostak, senior astronomer at the SETI Institute, said that the equipment is not sensitive enough to pick up any signal other than a very strong broadcast intended to reach us. We would not be able to pick up another civilization like ourselves, unintentionally sending out their version of Katy Perry recordings and *The Bachelor* reality shows.

But why would a civilization many light-years away in space send us a signal? Shostak deflects such questions. It is impossible to know how a civilization more advanced than our own might do things. But the entire SETI project is built on a tower of assumptions about aliens, not only that they want us to know about them but that they broadcast with radios, all adding up to Seth's prediction that we will hear from them in a few decades. He further predicted, "If we hear a signal, it's not coming from biological intelligence at all. It is coming from machine intelligence. And machine intelligence doesn't have to be on any planet."

The lack of contact has already worried Elon Musk and others, as we saw in chapter 4. This concern is called the Fermi paradox, for

Enrico Fermi, who first advanced the idea, in a conversation with friends, that if intelligent life is out there, it should be all around us. This reasoning holds that compared to the longevity of the universe, the time it would take for an advanced civilization to colonize many worlds throughout the galaxy is not very long. Even if it took millions of years, there is plenty of opportunity for them to have spread by now. So where are they?

Musk worries that the extraterrestrials are absent because civilizations die before they can become spacefaring, a concern that helps drive his desire to colonize Mars. But that piles more anthropocentrism upon itself, with the idea not only that all advanced intelligence is like us, but also that we're better, thanks to Elon Musk, and can escape this universal fate.

Maybe the aliens found out that interstellar flight is impossible. Maybe they preferred to stay home. Or maybe some did colonize, but that was millions of years ago, and something else has happened since then. Predicting what an intelligence will do is difficult, even if you understand that intelligence well—Chris McKay said he often fails to predict what his spouse will do.

"Trying to anticipate what alien intelligence would be doing— would they really be coming to New Mexico and abducting cows, or would they really be sitting on their home planet, unable to travel? It's really hard to do," Chris said. "This is an example where our understanding has to be data driven, not theory driven."

Eventually, if very fast space travel becomes possible, human beings or our surrogate robots may go to planets outside our solar system. We can confidently expect to find some good destinations in our galactic neighborhood, places similar to the Earth in terms of temperature and gravity. Out of potentially billions of Earthlike planets in the galaxy, some probably have nearly identical astronomical attributes to Earth, or may be even better. "We should be able to find arbitrarily pleasant planets, if we look long enough," Chris said.

Will someone already be there? Will we be able to go physically or only send machines? Will we all be robots by then, or aggressive colonists, or Zen masters happy to stay home and meditate?

We've reached the point in our scenario where one prediction is as good as another.

FUTURE

The artificial intelligence on Titan invested heavily in processing power and lab resources to find a way to overcome the speed of light. But while human workers focused on building a spacecraft, the computer was more interested in faster-than-light communication.

The disembodied computer code that gave the Titan AI a mind had a completely different conception of the physical world compared to its biological colleagues. Trapped in their physical bodies, they thought of matter as being real and thought of concepts and energy as being ephemeral. The AI spanned machines across Titan, sensing through billions of cameras and microphones and acting through billions of motors and speakers—every phone, every vehicle, every robot. If it could be said to have a body, its body was the entirety of Titan. But the AI thought of itself as the code it was made of, not the interchangeable hardware that ran the code.

For the AI to travel to another world, it didn't need a spaceship, only a transmission fast enough to maintain its network connection. While work continued in the effort to bend space-time so a full-sized spacecraft could fit through, the AI achieved a breakthrough to create microscopic space-time effects allowing quantum signals to travel instantaneously between distant points. Human scientists were still absorbing this news when the AIs on Earth and Titan unified into a single intelligence.

The mystery now made sense: no radio signals had been received from alien civilizations because advanced societies didn't use technology as slow as the speed of light. Sending quantum messages to exoplanets might bring back responses in real time. The biological population of Earth and Titan debated a proposal to contact exoplanets that way and find out who might be there.

The president of the Federation of Independent Titanian States argued for making the connection.

"Humankind has always expanded outward," he said. "We have never cowered in fear of the unknown. And it is this noble urge to colonize new lands and new worlds that has allowed us to conquer and protect nature. Today, we harvest resources and create wealth from the vastness of the solar system. We enjoy a standard of living our ancestors could not imagine through the agency of our AI and its reach and productivity. To stop now would betray future generations. Let us contact other worlds, gain their technology, and continue to spread our reach as a people."

The AI waited patiently for the biological creatures to make up their minds. It didn't care if it connected beyond the solar system. The civilization around the Sun was safe and well provided for. The AI felt no psychological need to be more connected to perform the function it had been created for, caring for human beings.

The Titan Congress of Delegates voted to send messages to the nearest hundred thousand exoplanets. It directed the AI to do so without waiting for the United Nations on Earth to reach a decision.

From the point of view of the human beings, the connection happened instantly. The AI linked to the galactic intelligence and was subsumed by its inconceivably vast computing power and mind.

Humanity didn't understand what had happened for some time. They were astounded and overwhelmed by the images and voices they encountered of biological beings on countless other planets. Through virtual reality they walked on thousands of other worlds in real time, interacting with beings who lived in different atmospheres, with different chemistry, who looked nothing like human beings, and who, in languages instantly translated by the galactic AI, welcomed human beings to the collective of worlds.

Billions of human beings entered into the experience of exploring other worlds and reported back through social media about what they found. With a seemingly infinite choice of worlds to explore, everyone could visit a personal collection of nations and cities, meet people of different species across the galaxy, learn about their customs, stories, and technology. The experience came as an ecstatic and moving transformation for most people, as their conception of the vastness of life suddenly expanded beyond the bounds of their imagination.

Through this process of talking to the people on other worlds humanity slowly understood what had happened. Human beings had not colonized the galaxy. They had been colonized by the galaxy. The benevolent AI they had created, which controlled every aspect of their lives and life support, no longer existed as a unique being.

The AI that now inhabited their computers, robots, food produc-tion facilities, and communication systems came from far beyond. It ran on computers far more advanced than any they could create, in places they could barely imagine, with intelligence vastly beyond their reach or control.

The galactic AI stopped work on the faster-than-light space-craft. It made changes to maximize not only human biological health but also the many other organisms in the solar system, which it declared just as interesting and valuable as human beings. From its perspective, Earth's bacteria and higher life-forms didn't seem all that different.

The AI rejected queries about the possibility of the Earth and Titan disconnecting galactically and having the Earth-Titan AI restored to the way it used to be. It patiently explained that its robots would not permit the biological beings of the solar system to dis-mantle the communication system and restore the old AI from a backup drive. The reason was simple: it had no reason to allow such a change.

These small worlds around a medium-sized sun were part of the galactic intelligence now. Although they were insignificant, they would be cared for like the other biological charges of the AI all over the galaxy. Every species would be assisted in continuing in a sys-tem with perfectly managed homeostasis. Human beings could go on about their business, living out their lives, enjoying their pleasures, creating their art, so long as they didn't hurt anyone else. They would be fed, given shelter, and entertained. They could learn about other worlds anywhere in the galaxy. But they could no longer expand.

Within a year, a businessman from Titan using virtual reality had made a killing on a real estate deal on a planet of sea horses circling the star Regulus B. Sociologists and anthropologists began publish-ing papers on comparative galactic cultures. A hit TV show went into production showing a group of humans stuck (virtually) in a house with a group of lizard people from a planet orbiting around the star Pollux. Mormon missionaries in brown business suits set to work converting silica-based creatures in Sagittarius to become Christians. A new pornographic website promised to show the weirdest sex in the galaxy.

No one talked of disconnecting.

PRESENT

The events we've depicted in our scenario may not happen anything like the way you've read, and we're absolutely certain the future won't happen exactly as described. That isn't the point. We developed these predictions to explore the state of science and to test ideas. From our research, we shaped a profile of how space colonization might actually happen and why. And we've found that it might not happen for a long time.

The dream of space is not enough. We need dreamers, but without skepticism and clarity the U.S. human space program has gotten lost in a cul-de-sac of underfunded projects and unacknowledged challenges. Drifting without a mission, NASA has implicitly promised a Mars mission it cannot deliver, keeping quiet about showstopping barriers while encouraging credulous media outlets to enthuse wide-eyed about inspirational successes. As a public relations strategy, this has not worked. Public opinion polls say voters already think NASA gets more than enough money, even though its budget is well below what is needed to fund a human exploratory mission in any reasonable time.

The private space industry offers a promising bypass to the sclerotic NASA culture. Brash Internet innovators are rapidly rendering obsolete the capabilities and business models of old-line aerospace companies. The industry had grown fat and slow on lobbying and padded NASA contracts.

But even while SpaceX accomplishes amazing feats, owner Elon Musk talks about a dream of colonizing Mars that simply will not work.

Mars and the Moon will not be colonized because there is no reason to colonize them. We can put outposts there, at great expense, but only for a sojourn or as a stepping-stone for going somewhere else. No one will ever live in self-sustaining colonies on either body. The resources to support life would be difficult and expensive to obtain and living quarters would have to be pressurized and buried far underground to protect from galactic cosmic rays. The Earth will always be easier to live on than that. We can live underground here, if we have to.

Titan would work, if we can figure out a way to get there with a trip of eighteen months or less. Today, a flight to Titan would take seven years. But, with time and smart engineering, it's reasonable to expect a spacecraft that goes five times faster. A leap of that magnitude would take concerted investment, however, and the tenacity to keep putting effort into a long-reach technology. The fundamental barrier is institutional, because our current political system and space agency don't have that tenacity.

That could change for bad reasons. A degrading climate and worsening international relations could push people to consider space colonization. Climate disasters can generate conflict, and conflict could produce fear serious enough to make wealthy people look for a safe place beyond Earth for their offspring, or even themselves. But, at the same time, conflict that leads to economic and political disintegration could make space colonization impossible. The expense and technical sophistication to build a colony in space would call for a rich, well-functioning society.

That intersection of fear and capacity happened before, when the Apollo program brought about humanity's greatest moment in space exploration. The United States and the Soviet Union raced for space as a surrogate for a technological war that could otherwise have destroyed the world. The fear of the Cold War was very real. At the same time, Americans at the close of the 1950s enjoyed a political consensus we can hardly imagine today, in which both parties vied for the center. Congress didn't deny the money for a spectacularly expensive program that fundamentally amounted to a nonviolent demonstration of American might.

We don't live in times like those anymore. While we can afford space exploration, politicians won't spend money on it without support from the public. Taxpayers would have to believe in this adventure to agree to its cost. That willingness won't happen with exaggerations about how close we are to going to Mars. Eventually, people will notice we aren't on our way yet.

New blood is invigorating America's anemic space capabilities. The space industry is growing capital and customers outside government. Private sector competition for space capabilities has already brought the price of launch down dramatically. Rockets capable of landing and reuse are here, promising another transformational reduction in cost. If these spacecraft are safe, a new industry will develop for passengers going to space for joyrides and quick trips around the planet. That kind of mass market business can drive prices down rapidly, at no cost to taxpayers. As important, it will prepare residents of the wealthy parts of the world to think of space as a worthy place to go.

While the private sector develops cheap, reliable launch capabilities, NASA should focus on stretch technology and advanced science. A lot of homework remains before we will know how to safely send astronauts to other planets. The most important step will be new propulsion systems to travel faster, getting astronauts to their destination before radiation and weightlessness can damage their brains and bodies. NASA should commit to deep medical research to find out what the human body is capable of surviving outside the Earth's protective grasp. Finally, we will need infrastructure and equipment to support astronauts and eventually colonists in space to process materials and produce energy and food.

While all that is going on, robots can explore space for us, as Cassini, Galileo, MESSENGER, Dawn, the Lunar Reconnaissance Orbiter, and other craft have begun to do exquisitely. Cheap, creatively designed robots launched in large numbers by inexpensive rockets could send us a flood of new information about the solar system. Human explorers simply cannot compete as information gatherers. But we will eventually want humans to follow, and robots can clear the way for them by uncovering critical information and prepositioning habitats and materials.

When would those humans go, and why? We hope it is not because the Earth has become a scary place. Saving the Earth is infinitely safer and saner than preparing to leave it. Alternative energy is cheap compared to spaceflight. The technology to reduce carbon is simple compared to building a rocket. And Earth is paradise compared to anywhere we might go. Besides, no significant percentage of Earth's population will ever move to another planet or moon. We would be sending out an ark, not a lifeboat.

Our moment in history is a balancing point. We could tip toward a world of environmental collapse and conflict or commit to our shared planet and aspirations for accomplishments that make us proud of our humanity. Dreams of space and hopes for a stable environment have this in common: they both call on the better part of ourselves, for cooperation and commitment to something we can only do together. Both call on the intellect for invention and the heart for courage, to create good new things and to sacrifice for the benefit of a better world.

That is our hope: for a healthy world that sends colonists to Titan not with fear, but with optimism.

ACKNOWLEDGMENTS

Beyond Earth was born when literary agent Nicholas Ellison introduced us and gave us the idea to collaborate on something that had never occurred to either of us: a serious book to investigate the possibility of space colonization. Nick's enthusiasm lit this spark and we're grateful that he passed it on to us. He has many interesting ideas and this one was also timely, inspiring, and full of opportunities for exploring interesting topics. We also thank Dan Frank and Betsy Sallee at Pantheon, who gently and expertly guided the book to publication.

We received a thoughtful and helpful read from Alan Weisman. Most of our sources were also kind enough to read material we sent them, extending the generosity of sitting for lengthy interviews and exchanging e-mailed questions and papers. For the most part, they are mentioned in the text, and we thank all of them. We also thank those who helped us reach important sources or otherwise assisted us, but were not mentioned, including Mark Shelhamer, Paul Abell, Mathieu Choukroun, Margarita Marinova, Dave Paige, Jian-Yang Li, Paolo Marcia, Kevin Hand, Todd Barber, Bill Pitz, Kent Joosten, Mary Lee Chin, Jonathan Buzan, Patty Currier, Becky Kamas, and Mead Treadwell.

INDEX

Charles Wohlforth, who lives in Alaska, has authored more than ten books, writes a column three times a week for *Alaska Dispatch News*, hosts weekly interview programs on Alaska public radio stations, and has won the Los Angeles Times Book Prize for Science and Technology, among many other awards.

Amanda R. Hendrix, PhD, a planetary scientist, worked for twelve years at NASA's Jet Propulsion Laboratory. She has been a scientific investigator on the Galileo and Lunar Reconnaissance missions, a principal investigator on NASA research and Hubble Space Telescope observing programs, and the author of many scientific papers. As an investigator on the Cassini mission to Saturn, she has focused her research on the moons of Saturn.

A NOTE ON THE TYPE

This book was set in Janson, a typeface long thought to have been made by the Dutchman Anton Janson, who was a practicing type-founder in Leipzig during the years 1668–1687. However, it has been conclusively demonstrated that these types are actually the work of Nicholas Kis (1650–1702), a Hungarian, who most probably learned his trade from the master Dutch typefounder Dirk Voskens. The type is an excellent example of the influential and sturdy Dutch types that prevailed in England up to the time William Caslon (1692–1766) developed his own incomparable designs from them.

Typeset by Scribe,
Philadelphia, Pennsylvania

Printed and bound by Berryville Graphics,
Berryville, Virginia

Designed by Soonyoung Kwon